EMPLOYEE'S RECEIPT

I acknowledge receipt of Keller's Official OSHA Construction Safety Handbook, which covers 21 different safety topics. These topics include the following:

Confined Space Entry (1926.21, .57, .103, .154, .650-.651, .350-.354, .800)

Electrical Safety (1926.400-.449)

Emergency Response (1926.24, .35, .64, .65, .150-.155)

Ergonomics and Back Safety

Excavations (1926.650-.652)

Fall Protection (1926.500-.503)

First Aid & Bloodborne Pathogens (1926.50, 1910.1030)

Hazard Communication (1926.59)

Health and Wellness

Jobsite Exposures (1926.53, .54, .55, .57, 60, .62, .1101-.1152)

Ladders & Stairways (1926.1050-.1060)

Lockout/Tagout (1926.417, .702)

Materials Handling & Storage (1926.20, .250-.252, .550-.555, .600-.606, .1000-.1003)

Motor Vehicles (1926.20, .600,-.601)

Personal Protective Equipment
    Eye Protection (1926.102)
    Foot Protection (1926.96)
    Hand Protection
    Head Protection (1926.100)
    Hearing Protection (1926.101)
    Respiratory Protection (1926.103)

Scaffolds (1926.451-.454)

Site Safety & Security (1926.21, .25, .200-.203, 250, .252)

Slips, Trips, & Falls

Tool Safety (1926.300-.307)

Welding, Cutting, and Brazing (1926.350-.354)

Work Zone Safety (1926.200-.203)

_____
Employee's Signature   Date

_____
Company

_____
Company Supervisor's Signature

NOTE: This receipt shall be read and signed by the employee. A responsible company supervisor shall countersign the receipt and place it in the employee's training file.

REMOVABLE PAGE - PULL SLOWLY FROM TOP RIGHT CORNER

W9-ASQ-253

Keller's

OFFICIAL OSHA CONSTRUCTION SAFETY HANDBOOK

©Copyright 2005

J. J. Keller & Associates, Inc.

3003 W. Breezewood Lane, P. O. Box 368

Neenah, Wisconsin 54957-0368

Phone: (920) 722-2848

www.jjkeller.com

Library of Congress Catalog Card Number: 2004107866

ISBN 1-59042-494-8

Canadian Goods and Services Tax (GST) Number: R123-317687

Printed in the U.S.A.

Fifth Edition, First Printing
January, 2005

TABLE OF CONTENTS

Introduction

Introduction

Your employer cares about your safety and health. *Keller's Official OSHA Construction Safety Handbook* is just one part of your company's program to teach you about job hazards you face every day.

Occupational Safety and Health Administration (OSHA) regulations often require training on particular topics such as fall protection or personal protective equipment (PPE). This handbook covers 21 topics, including subjects like hazard communication and excavations. It serves as an aid to your supervisor or instructor in training you on your company's safety procedures.

OSHA regulations for construction are in the Code of Federal Regulations, Title 29, Part 1926. Individual regulations will be identified throughout this handbook as 29 CFR 1926.(section number).

This handbook covers subject areas that are the leading cause of injury and death at construction sites. Falls account for 33 percent of construction fatalities, highway crashes 13 percent, caught in between hazards such as trenching 10 percent, and electrical accidents 15 percent.

This book is yours to keep. Use it to follow along with your instructor during training sessions, then keep it to use as a handy reference after you finish your safety training program. You can look up safety and health facts and find answers to safety questions that come up while you're working. Your handbook can be kept in your locker or toolbox, keeping safety information at hand whenever you need it.

Chapter reviews found at the end of each chapter are to be used to review key points and main ideas. Fill out the review when instructed to do so by your instructor. Only one answer is correct for each question. Circle or underline the letter next to the correct answer, as instructed to do so by your instructor. The reviews are perforated so that you can tear them out easily and turn them in.

You, together with your employer, can make every workday a safer one by following the guidelines in *Keller's Official OSHA Construction Safety Handbook.*

CONFINED SPACE ENTRY: GET IN AND OUT SAFELY

Each day thousands of construction workers are exposed to possible injury or death in what OSHA calls "confined or enclosed spaces."

During a five-year period from 1997 to 2001 BLS investigated 458 fatal confined space deaths. Of that total, 89 deaths were construction workers. Don't become one of the statistics. It only takes a little time to properly prepare to enter a confined or enclosed space.

What Is a Confined or Enclosed Space?

The OSHA construction regulations define a confined or enclosed space as any space having a limited means of getting out, and which can gather toxic or flammable gases, vapors, or has oxygen-deficient air. NIOSH also says a confined space is one that has unfavorable natural ventilation.

Where Are the Regulations?

Regulations governing entry into confined or enclosed spaces are specified by OSHA. However, for construction, they are not neatly defined in one area as they are in the general industry rules (29 CFR 1910.146). They are located in various spots in the construction regulations, and cover such topics as training, hazards, precautions, personal protective equipment, ventilation, and excavations.

Examples of confined or enclosed spaces include storage tanks, process vessels, bins, boilers, ventilation and exhaust ducts, sewers, underground utility vaults, excavations, manholes, tunnels, pipelines, and open top spaces more than four feet in depth, such as pits, tubs, vaults, and vessels.

Confined or Enclosed Space Hazards

There are many situations and hazards that can cause a confined space to become deadly. Materials being used such as cleaning or bonding liquids, work being done such as welding, or the effects of the environment can cause dangerous vapors, gases, and mists to accumulate in these spaces. The result can be fires, explosions, and physical hazards.

Physical Hazards

The largest number of deaths in confined spaces are caused by atmospheric problems. However, if trench cave-in deaths are included as confined-space-related deaths, then physical hazards would be the largest group.

Physical hazards include:

- **Hazardous energy**—Activated electrical, mechanical, and hydraulic energy can cause injury in a confined space; therefore, it must be deenergized and locked out before you go to work in that space.

- **Cave-in**—When cave-ins are possible, OSHA's excavation rules must be followed.

- **Drowning**—Heavy rain or water from pipes can enter the space.

- **Underground utilities**—Lines containing steam, gases, or coolants should be shut off.

- **Communication problems**—Poor communication systems may delay rescue.

- **Heat**—Temperatures can build up quickly in a confined space and cause exhaustion or dizziness.

- **Noise**—Sound from equipment and workers reverberates in the space and can make it difficult to hear important directions or warnings.

- **Entry and exit difficulties**—Entry and exit openings can be limited by size or location.

Atmospheric Hazards

Asphyxiation caused by atmospheric problems is the main hazard in confined spaces.

Oxygen Deficiency

Most confined space accidents are related to atmospheric conditions inside the space and the failure to continuously monitor the air and ventilate as necessary. In general, the primary risk associated with confined spaces is oxygen deficiency.

Normal air contains 20.8 percent oxygen. The minimum safe level, an OSHA requirement, is 19.5 percent. OSHA also says that the maximum safe level is 23.5 percent. At 16 percent you will feel disoriented and between eight percent and 12 percent, you will generally become unconscious. If the air has too much oxygen (over 23.5 percent) it is considered oxygen rich and becomes an explosion or fire hazard.

Oxygen is reduced in a space when it is replaced by another gas or used up. It can also be displaced by other gases such as argon, nitrogen, or methane. Oxygen can be consumed during combustion of flammable substances, as in welding, cutting, or brazing. A more subtle form of consumption of oxygen occurs during bacterial action, as in the fermentation process. Oxygen may also be consumed during slow chemical reactions such as the formation of rust.

Flammable Air

Fire and explosion are serious dangers in a confined space. Fumes and vapors will ignite more quickly in the trapped air.

Flammable and combustible gases or vapors may be present from previous contents, tank coatings and preservatives, and welding operations. In locations where flammable vapors may be present, precautions must be taken to prevent ignition by eliminating or controlling the source of ignition or eliminating the flammable air before working. Sources of ignition may include smoking, cutting and welding, hot surfaces, and frictional heat.

Toxic Air Contaminants

Toxic air contaminants come from material previously stored in the confined space or as a result of the use of coatings, cleaning solvents, or preservatives. The work being performed in a confined space could also give off a toxic gas. An example of this would be a welding operation that gives off carbon monoxide and oxides of nitrogen and ozone.

Unfortunately, you cannot see or smell most toxics, but they present two types of risk in a confined space: they can irritate your respiratory or nervous system; or some toxic chemicals can cut off your oxygen supply, get into your lungs, and asphyxiate you.

CONFINED SPACE

Working in Confined Spaces

If you are required to enter a confined space, your supervisor is required to instruct you as to:

- What kinds of hazards you may run into and why those hazards are dangerous.

- The necessary precautions to take for each type of hazard.

- The use of any protective and/or emergency equipment and instruments required.

Although construction people are not required to follow the permit-required confined space entry program that those working in general industry are, many of the requirements of the construction regulations fit nicely into the program. And, it is a way to maximize your safety. Your participation in the development and implementation of a permit-required confined space entry program is encouraged.

Before You Enter a Confined Space

Before you enter a confined or enclosed space there are certain procedures you must follow to ensure the space is safe.

Obtain an Entry Permit

Although this is not required by OSHA for construction, it is a good idea to obtain a written authorization (entry permit), signed by your supervisor, prior to you entering a confined space. This would be an excellent time for you and your jobsite supervisor to discuss the necessary precautions for the job you are doing.

Control Hazardous Energy

Use lockouts and tags to prevent accidental startup of equipment while you are working in the confined space. Cut off steam, water, gas, or power lines that enter the confined space. Use only safe, grounded, explosion-proof equipment and fans.

When you are working in a confined or enclosed space that has exposed energized parts, your employer must provide some type of protective insulated shield to prevent contact with these parts.

Test the Air

Your company should make it a practice to test the air every time you are required to go into a confined space. Adequate precautions must be taken to prevent your exposure to:

- Air containing less than 19.5 percent oxygen.

- A concentration of a flammable gas in excess of 20 percent of the lower flammable limit of the gas.

- Any other toxic hazardous air.

Your company should have special instruments for testing the levels of oxygen, combustibility, and toxicity in confined spaces.

Excavations over four feet deep where oxygen deficiency or other hazardous atmosphere exists or could reasonably be expected to exist must be tested.

Test the air prior to going into the confined space and on a regular basis during your stay for presence of sufficient oxygen and absence of hazardous levels of toxic or combustible gases. Once the space is opened, test the air from top to bottom. Some gases like propane and butane are heavy, and they will sink to the bottom of the space. Light gases like methane will rise to the top. So you need to be sure to check all levels.

After you are sure that the oxygen level is adequate and there is nothing combustible in the space, test for toxicity.

If you are assigned to enter a confined space, you or your authorized representative have the right to watch all air monitoring and see all monitoring results and written certifications that the space is safe to enter.

If tests indicate the space is unsafe to enter notify your supervisor. Your company should have a tag out system to mark the confined space UNSAFE FOR ENTRY. The tag must remain in place until tests indicate you can safely enter the space.

When toxic substances are present for which no equipment to test the atmosphere is available, you must be permitted to enter the confined space only with the use of supplied air respiratory equipment and other appropriate personal protective equipment.

Equipment

When you enter a confined space through a manhole or other small opening you should have means to be quickly removed in case of an emergency. You should use a chest harness or a full body harness with retrieval line attached at the center of your back near the shoulders. Body belts used for retrieval can cause the person to get stuck in small exit openings.

Ventilation

When you are welding, cutting, or heating in a confined space, ventilation must be provided to ensure oxygen levels are safe and toxic or flammable air is not at dangerous levels.

When sufficient ventilation cannot be obtained without the ventilation equipment blocking your means of escape, you must be provided an air line respirator and trained to use it properly.

Use ventilating equipment where possible. Ventilation should maintain an oxygen level between 19.5 percent and 23.5 percent. It also should keep toxic gases and vapors to within accepted levels prescribed by OSHA.

Rescue Procedures

When workers enter a confined space, at least one person should remain outside to summon help or offer assistance. Your company must have written emergency rescue procedures that require trained personnel to be available and stationed where they may reach victim(s) within a time frame appropriate for the hazards of the confined space.

The trained attendant should be knowledgeable in first aid and cardiopulmonary resuscitation (CPR). The attendant also must maintain constant communication with those inside the space either visually, by radio, or by field telephone. If a situation arises that requires emergency entry, the attendant should not enter until additional help arrives.

A rope tied around a worker's waist is not an acceptable rescue method. It does not allow a single attendant to pull an injured worker out of a space. A full body harness and lifeline is a better approach. It can be attached to a block and tackle that a single rescuer can operate.

Emergency rescue equipment such as a self-contained breathing apparatus (SCBA), a safety harness and line, or a basket stretcher, must be readily available where hazardous atmospheric conditions exist or may reasonably be expected to develop during work in an excavation. The equipment must be attended when in use.

Work at Working Safely

Sometimes the confined space you are entering will not appear to be hazardous. It may have been entered on the last shift with no problems, and may not show signs of being dangerous. At other times there may be indications of danger—the distinct odor of toxic atmospheres, arcing of electrical equipment, or the presence of loose material. Recognition of these dangers is critical and must be a part of your company's safety program.

Employee _____

Instructor_____

Date _____

Company _____

CONFINED SPACE ENTRY REVIEW

1. A confined space:
   a. Allows for continuous human occupancy.
   b. Has restricted entry and exit.
   c. Is large enough and configured to allow workers to enter and perform required duties.
   d. Both b and c.

2. Confined space hazards include:
   a. Engulfment by materials in the confined space.
   b. Hazardous atmospheres such as oxygen deficiency or enrichment.
   c. Toxic or flammable contaminants.
   d. All of the above.

3. Excavations over _____ feet deep where an oxygen deficiency or other hazardous atmospheres exist, or could be expected to exist, must be tested.
   a. six
   b. five
   c. four
   d. three

4. Adverse conditions that may affect a confined space entrant are:
   a. Heat.
   b. Noise.
   c. Working in a cramped position.
   d. All of the above.

5. The oxygen content of oxygen enriched air is:
   a. Less than 21 percent.
   b. Greater than 21 percent.
   c. Less than 19 percent.
   d. Greater than 23.5 percent.

6. The oxygen content of oxygen deficient air is:
   a. Greater than 15 percent.
   b. Less than 19.5 percent.
   c. Less than 11.5 percent.
   d. Less than 15 percent.

7. If a confined space has a hazardous atmosphere:
   a. Test the air before entering and as often as necessary.
   b. Ventilate the space.
   c. Use an appropriate respirator, if necessary.
   d. All of the above.

8. Wearing a full body harness and retrieval line:
   a. Makes a non-entry rescue possible.
   b. Is only required during entries into vertical spaces more than five feet deep.
   c. Is always required, even if it would increase the hazards of the entry.
   d. Is not required if the attendant will be able to see the entrant at all times during the entry.

9. Which of the following is *not* appropriate retrieval equipment for a non-entry rescue?
   a. Full body harness with retrieval line attached at the center of entrant's back near the shoulders.
   b. Wristlets attached to retrieval line when full body harness creates a greater hazard.
   c. A body belt with a retrieval line when full body harness creates a greater hazard.
   d. Chest harness with retrieval line attached at the center of entrant's back near the shoulders.

10. Test the air in a confined space for proper oxygen content:
    a. Before you enter.
    b. Regularly during your stay.
    c. From the top level to the bottom level.
    d. All the above.

CONFINED SPACE

ELECTRICAL SAFETY: A SHOCKING HAZARD

Exposed junction box wiring, damaged extension cords, and temporary set-ups are some of the electrical hazards construction workers face daily. And electricity can be deadly, exposing you to such dangers as shock, electrocution, fires, and explosions.

In 2002, according to the Bureau of Labor Statistics, 2,967 construction workers were injured after coming into contact with electric current. There were also 289 deaths from electrocution. In addition, OSHA's report, *An Analysis of Fatal Events in the Construction Industry 2002*, revealed that of the 719 fatal events studied, 110 deaths (or 15.4 percent) resulted from contact with electric current.

Where Are the Regulations?

The OSHA electrical regulations are located in Subpart K of 29 CFR 1926. These electrical safety rules are divided into: safety requirements for installing and using equipment, safety-related work practices, safety-related maintenance and environmental issues, and safety requirements for special equipment. Your employer is required to provide you with training in safe electrical work practices, and has a duty to provide hazard-free equipment and work situations.

How Does Electricity Work?

To handle electricity safely, you need to understand how it acts, how it can be directed, the hazards it presents, and how those hazards can be controlled. When you turn on your circular saw or throw a circuit breaker, you allow current to flow from the generating source, through conductors (wiring), to the area of demand or load (equipment or lighting).

A complete circuit is necessary for the flow of electricity through a conductor. A complete circuit is made up of a source of electricity, a conductor, and a consuming device (load) such as a portable drill.

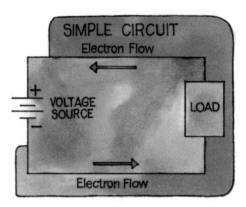

Volts = Current X Resistance (or V=IR) is an equation known as Ohm's Law. The equation shows the relationship between three factors. This relationship makes it possible to change the qualities of an electrical current but keep an equivalent amount of power.

A force or pressure must be present before water will flow through a pipeline. Similarly, electrons flow through a conductor because an electromotive force (EMF) is exerted. The unit of measure for EMF is the volt.

For electrons to move in a particular direction, a potential difference must exist between two points of the EMF source. The continuous movement of electrons past a given point is known as current. It is measured in amperes.

The movement of electrons along a conductor meets with some opposition. This opposition is known as resistance. Resistance to the flow of electricity is measured in ohms. The amount of resistance provided by different materials varies. For example, most metals offer little resistance to the passage of electric current. However, glass, mica, rubber, plastic, or wood, have a very high resistance to the flow of electricity.

What Are the Hazards of Electricity?

Electricity's primary hazards are shock and possible electrocution, burns, arc-blast, explosions, and fires.

Shock

Electric currents travel in closed circuits. You get a shock when some part of your body becomes part of an electric circuit. An electric current enters the body at one point and exits the body at another. You will get a shock if you touch:

- Both wires of an electric circuit.

- One wire of an energized circuit and ground.

- A metallic part that is "hot" because it is contacting an energized wire and you are in contact with the ground.

A shock's severity depends on several factors:

- How much electric current flows through your body (measured in amperes).

- What path the electric current takes through the body.

- How long the body is part of the electric circuit.

The effects of an electric shock on the body can range from a tingle to immediate cardiac arrest. Low voltages can be just as deadly as high voltages if the body is a part of the circuit longer.

LOW VOLTAGE DOES NOT IMPLY LOW HAZARD!

Water and Electric Shock

Water presents an interesting and potentially dangerous situation. In its pure state, water is a poor conductor of electricity. However, if even small amounts of impurities are present in the water (salt and acid in perspiration, for example), it becomes a ready electrical conductor.

Therefore, if water is present at your worksite, or on your skin, be extra careful around any source of electricity. Carelessness with the combination of water and electricity could cost you your life.

Burns and Other Injuries

Electric burns are one of the most serious injuries you can receive and should be given immediate attention. A severe shock can also cause considerably more damage to the body than is visible. For example, a person can suffer internal bleeding and severe destruction of tissues, muscles, nerves, and internal organs. This is the result of current flowing through tissue or bone, generating heat, and causing injury. In addition, shock is often only the beginning in a chain of events. The final injury may well be from a fall, cuts, burns, or broken bones.

Arcing and Sparking

Arcing or sparking occurs when high-amperage currents jump from one conductor to another through air, generally during opening or closing circuits, or when static electricity is discharged. Fire may occur if the arcing takes place in an atmosphere that contains an explosive mixture. Also, the arc could start other flammable material on fire.

Explosions

Explosions occur when electricity provides a source of ignition for an explosive mixture in the atmosphere. Ignition can be due to overheated conductors or equipment, or normal arcing at switch contacts. OSHA standards, the *National Electrical Code*, and related safety standards have precise requirements for electrical systems and equipment when applied in such areas. Your employer is required to make a hazard assessment and give you instructions in these cases.

Fires

Electricity is one of the most common causes of fire. High resistance connections, a primary source of ignition, occur where wires are improperly spliced or connected to other components such as receptacle outlets and switches.

Heat develops in a conductor from current flow. If you put more current through a conductor than it can handle, it may get hot enough to start a fire.

Causes of Electrical Accidents

When working with electricity, accidents and injuries are caused by one or a combination of the following factors:

- Unsafe equipment and/or installation.

- Unsafe workplaces caused by environmental factors.

- Unsafe work practices.

Report instances of unsafe equipment, equipment installations, and unsafe worksites to your employer. Practice safe work habits.

Preventing Electrical Accidents

Protection from electrical hazards is one way to prevent accidents caused by electric current. Protective methods to control electrical hazards include insulation, electrical protective devices, guarding, grounding, personal protective equipment (PPE), and safe work practices.

Insulation

Insulation keeps conductors from being exposed. While OSHA requires that insulation be suitable for the voltage and conditions under which the item will be used such as temperature, moisture level, and fumes, it is in your best interest to check your equipment for insulation breakdown. Check for exposed wires, scuffed insulation on extension cords, for wires wearing through, and broken wiring. Non-conducting coatings on tool handles also aid in insulating from electrical shock.

Circuit Protective Devices

Circuit protective devices, including fuses, circuit breakers, and ground-fault circuit-interrupters (GFCIs), are critically important to electrical safety. These devices are designed to automatically limit or shut off the flow of electricity in the event of a ground-fault, overload, or short circuit in a wiring system.

Fuses and circuit-breakers are over-current devices that are placed in circuits to monitor the amount of current the circuit is carrying. They automatically open or break the circuit when current flow becomes excessive and therefore unsafe.

Fuses and circuit breakers are used to protect conductors and equipment. They prevent overheating of wires and components that might otherwise create hazards for workers. They also open the circuit under certain hazardous ground-fault conditions.

However, the only electrical protective device whose sole purpose is to protect people is the ground-fault circuit-interrupter. The GFCI is not an overcurrent device. It senses an imbalance in current flow over the normal path and opens the circuit in a fraction of a second.

Although the GFCI does not protect you from line-to-line hazards (holding two hot or one hot and one neutral wire), it does provide protection against the most common form of electrical hazard for construction workers—the ground fault.

While most portable electric tools have an equipment grounding conductor and many are double insulated, these methods are not 100 percent safe. A grounding wire could break or a cord could become defective. Using a GFCI overcomes these insulation problems.

Guarding

Any "live" parts of electrical equipment operating at 50 volts or more must be guarded to avoid accidental contact. This protection can be accomplished by installing equipment:

- In a room, enclosure, or vault;

- Behind substantial screens, cages or partitions;

- On a balcony, platform, or elevated gallery area; or

- At least eight feet above the floor of a work area.

Any entrance to an area containing "live" parts of electrical equipment must be marked with warning signs. These signs should forbid entrance except by qualified persons.

Grounding

Grounding is required to protect you from electrical shock, safeguard against fire, and protect against damage to electrical equipment. There are two kinds of grounding:

- Electrical circuit or system grounding, accomplished when one conductor of the circuit is intentionally connected to earth. This protects the circuit if lightning strikes or other high voltage contact occurs. Grounding a system also stabilizes the voltage in the system so "expected voltage levels" are not exceeded under normal conditions.

- Electrical equipment grounding occurs when the equipment grounding conductor provides a path for dangerous fault current to return to the system ground at the supply source of the circuit should the insulation fail.

When electrical equipment is grounded, a low-resistance path is intentionally created to earth. This path has enough current-carrying capacity to prevent any buildup of voltages in the equipment.

Grounding does not guarantee that you will never receive a shock. Be sure any equipment you work on is properly grounded and that you do not defeat grounding devices (such as the grounded plugs of portable power equipment).

Assured Equipment Grounding Conductor Program

The assured equipment grounding conductor program is an inspection program covering:

- All cord sets (extension cords).

- Receptacles that are not a part of the permanent wiring of the structure.

- Equipment connected by cord and plug that is available for use or is used by employees.

This inspection program includes electrical equipment that must be visually inspected for damage or defects before each day's use. Any damaged or defective equipment must not be used until repaired.

Under this program, OSHA requires the following two tests to be performed before the first use of new equipment, after suspected damage to equipment, and at three month intervals:

- A continuity test to ensure that the equipment grounding conductor is electrically continuous.

   The test must be performed on receptacles that are not part of the permanent wiring of the building or structure, on all cord sets, and on cord-and plug-connected equipment that is required to be grounded.

- A test to ensure that the equipment grounding conductor is connected to its proper terminal.

   This test must be performed on receptacles and plugs.

Personal Protective Equipment

If you work in an area where there are potential electrical hazards, your employer must provide you with protective equipment. You must use electrical protective equipment (see 29 CFR 1926 Subpart E) appropriate for the body parts that need protection, and for the work to be done. An example of this would be the OSHA requirement to wear insulated, nonconductive gloves when using a jack-hammer if striking underground powerlines is a possibility.

Safe Work Practices for Working with Electricity

If your job requires you to work with electrical equipment, you need to have a healthy respect for its power. In general, you should be sure that any tools you use are in good repair, that you use good judgment when working near electrical lines, and that you use appropriate protective equipment.

Lockout/Tagout

Having electrical current unexpectedly present when you are working on a piece of equipment is no joke! Before an authorized person begins any repair work or inspection of electrical equipment, the current should be turned off at the switch box, and the switch padlocked in the OFF position. You must also tag the switch or controls. The tag should indicate which circuits  or pieces of equipment are out of service.

General Precautions

The following general rules apply to all work at your jobsite:

- Maintain your electrical equipment according to manufacturer and company standards.

- Respect warning signs, fences, cages or other barriers for special electrical hazards.

- Regularly inspect tools, cords, grounds, and accessories before starting work every day.

- Repair only those items that you are authorized to repair. If you are not qualified, arrange to have equipment repaired or replaced immediately.

- Use safety features like three-prong plugs, double-insulated tools, and safety switches. Keep machine guards in place and follow proper procedures.

- Install or repair equipment only if you're qualified and authorized to do so. A faulty job may cause a fire or seriously injure you or other workers.

- Keep electric cables and cords to equipment clean and free from kinks. Never carry equipment by the cords.

- Extension cords are more vulnerable to damage; use and maintain them properly. Never:

  - Use worn or frayed cords.

  - Fasten with staples, hang from nails, or suspend by wire, or any other method that could damage the insulation.

  - Run them through holes in walls, ceilings, floors, doorways, or windows without protection.

- Don't touch water, damp surfaces, ungrounded metal, or any bare wires if you are not protected. Wear approved rubber gloves when working with live wires or ungrounded surfaces, and rubber-soled shoes or boots when working on damp or wet surfaces.

- Don't wear metal objects (rings, watches, etc.) when working with electricity. They might make you a ground and could cause injury.

- If you are working near overhead power lines of 50 kiloVolts (kV) or less, you or any equipment you are using must not come any closer than ten feet from the lines. Add four inches of distance for every 10 kV over 50 kV.

Work at Working Safely

Safety should be foremost in your mind when working with electrical equipment. Because you face hazards from jobsite conditions, your tools, and the electricity that powers them, wear protective equipment whenever it's specified, use all safety procedures, and work with tools correctly. Never let overconfidence lead to taking unnecessary risks. If you're not sure —don't touch.

NOTES

Employee _____

Instructor_____

Date _____

Company _____

ELECTRICAL SAFETY REVIEW

1. A _____ is necessary for the flow of electricity through a conductor.
    a. Insulated circuit.
    b. Loop.
    c. Complete circuit.
    d. Consuming device.

2. The unit of measure for electromotive force is the _____.
    a. Amp.
    b. Resistance.
    c. Ohm.
    d. Volt.

3. A complete source is made up of a source of electricity, a _____, and a consuming device (like a portable drill).
    a. Load.
    b. Electron.
    c. Conductor.
    d. Circuit.

4. The severity of an electric shock depends on:
    a. The path the current takes through the body.
    b. How long the body is part of the electric circuit.
    c. The thickness of insulation on the wire.
    d. Both a. and b.

5. One way a shock can occur is:
    a. When energized parts operate at more than 204 volts, nominal.
    b. If you touch both wires of an electric circuit.
    c. When you touch a deenergized conductor and ground.
    d. When you touch a "dead" metallic part while you are grounded.

6. Circuit protective devices automatically limit or shut off current flow during:
   a. A ground fault.
   b. An overload.
   c. A short circuit.
   d. All of the above.

7. The two types of grounding are:
   a. Circuit and equipment.
   b. Direct and alternating.
   c. Supply and return.
   d. Positive and negative.

8. Fuses and circuit breakers are examples of:
   a. Grounding equipment.
   b. Circuit protective devices.
   c. Electrical equipment guards.
   d. Electrical insulation.

9. Live parts of electrical equipment operating at 50 volts for more:
   a. Must be locked out.
   b. Must have hazard alert labels.
   c. Must be guarded to avoid accidental contact.
   d. All of the above.

10. The assured equipment grounding conductor program covers inspection of:
    a. All extension cords.
    b. Receptacles that are not a part of the structure's permanent wiring.
    c. Cord- and plug-connected equipment.
    d. All of the above.

EMERGENCY RESPONSE: DEALING WITH SITE INCIDENTS

Despite all efforts to provide a safe and healthy worksite, accidents still happen. Weather and other emergencies, unrelated to your job, can also cause problems. However, proper planning can save lives and minimize property damage during an emergency. Effective safety and health programs should have Emergency Action Plans (EAPs) for emergencies such as personal injury, fires, severe weather, accidental releases of toxic gases, chemical spills, or explosions.

Where Are the Regulations?

OSHA has issued a number of regulations covering emergency planning and training for construction. These include:

29 CFR 1926	Name
.24	Fire protection and prevention
.35	Employee emergency action plans
.64	Process safety mgmt. of highly hazardous chemicals
.65	Hazardous Waste Operations and Emergency Response (HAZWOPER)
.150-.155	Fire Protection and Prevention

OSHA doesn't require formal action plans for tornadoes, earthquakes, or other severe weather conditions. However, your company should have an EAP in case these situations occur.

Elements of an Emergency Action Plan

Your company's EAPs should include the following elements:

- Best way to report fires and other emergencies,

- How the alarm system sounds for different emergencies,

- Emergency escape procedures and routes to take,

- Procedures for workers who stay behind to operate critical equipment or functions before they evacuate,

- Head count procedures after an evacuation is completed,

- Rescue and medical duties for designated employees, and

- Names or job titles of persons or departments to be contacted for further information regarding duties under the plan.

The plans must be kept at each worksite and made available to all employees. If your company has ten or fewer employees, the EAPs do not have to be written. Instead, your supervisor can give them to you verbally.

Training for Emergencies

Training must be given to you:

- When first starting your job,

- Whenever your job responsibilities change, and

- Whenever the EAPs are first developed or changed.

Your employer will designate and train certain employees to have additional responsibilities to facilitate a safe and orderly evacuation.

Specific Emergency Action Plans

Process Safety Management

If your company is involved in maintenance or repair, major renovation, or specialty work in a facility where exposure to highly hazardous chemicals (such as a petrochemical plant) is possible, an EAP is required. Your employer is responsible for training you on the known potential hazards at that specific site and the EAP you should follow.

Fire Protection/Prevention

Because the best defense against a fire is to prevent it from starting, your employer must have a plan for fire protection and prevention at the jobsite. Being aware of the hazards and causes of fires can help you prevent them and protect yourself. Also, you need to know what to do if a fire starts.

Fire is a reaction characterized by the light and heat of combustion. It has four basic components: fuel, heat or an ignition source, oxygen, and a chain reaction (the process of combustion). To burn, a fire needs enough oxygen to sustain combustion, enough heat to raise the combustible material to its ignition temperature, and some sort of fuel to feed the chain reaction.

Although many materials stored at a jobsite are not flammable, other materials commonly found there are potential "fuels." These include wood products such as lumber, flammable liquids such as gasoline and propane, and a variety of chemicals.

It doesn't take very long for a combustible environment to turn into a fire. A few minutes can be the difference between a narrow escape and an emergency. Knowing how to recognize conditions that can lead to a fire can help prevent them.

Fire Hazards

There are two major jobsite fire hazards you should be aware of—flammable and combustible liquids and heating devices.

Flammable and Combustible Liquids

Flammable liquids give off ignitable vapors. The combination of these vapors with ignition sources (such as a hand tool that sparks or a cutting torch) at construction sites is a serious threat. Store and use flammable and combustible liquids properly.

Only use approved containers and portable tanks for storing and handling flammable and combustible liquids. The container must be red, and if the flammable liquid has a flashpoint at or below 80°F, it must have a yellow band or the name of the dangerous liquid stenciled in yellow.

Heating Devices

Many fires have started because heating devices were used improperly or in the wrong conditions or environments. In addition, some temporary heating devices have fume and direct burn hazards associated with them. Because of these fire hazards, you should:

- Use adequate ventilation to reduce your exposure to dangerous fumes,

- Keep the heater sufficiently clear of combustible material, including wood floors,

- Position heaters on stable, level surfaces,

- Not use solid-fuel salamanders, and

- Make sure that oil-fired heaters have a primary safety control to stop the flow of fuel in the event of flame failure.

Fire Response

Because of the deadly danger of fire, you should know how to size up a fire and how to respond in a fire emergency. If you are faced with a fire emergency, it's important to protect yourself and other employees.

Fire Extinguishers

Only properly trained employees should use fire extinguishers. If you use an extinguisher, be sure to use one that is designed for that type of fire. Using the wrong extinguishing agent on a fire may increase the intensity of the fire.

The National Fire Protection Association (NFPA) has classified fire into four general types based on the combustible materials involved and the kind of extinguisher needed to put them out. The four fire classifications are A, B, C, and D. Each classification has a special symbol and color identification. Look at your fire extinguishers—classifications should be found on their labels as follows:

Class	Materials burning	Extinguishing agent	Symbol
A	Wood, paper, rubber, plastics	Water, dry chemicals	△
B	Flammable liquids, gases, greases	Carbon dioxide, dry chemicals	❏
C	Electrical equipment, wiring, fuse boxes, circuit breakers, machinery	Carbon dioxide, dry chemicals	○
D	Combustible metals	Special techniques, do not use common extinguishers	☆

There is also a Class K extinguisher that is used for kitchen fires.

If the fire can be contained or extinguished, a properly trained person should follow the "PASS" method of extinguisher use. This method includes holding the extinguisher upright, and:

- **P**ulling the pin, standing back eight or ten feet,

- **A**iming the extinguisher at the base of the fire,

- **S**queezing the handle, and

- **S**weeping at the base of the fire with the extinguishing agent.

If you aim high at the flames, you won't put out the fire. Remember, too, that most extinguishers have a very limited operation time, only 8-10 seconds, so you have to act fast and spray correctly at the base of the fire, not at smoke or flames.

Hazardous Waste Operations and Emergency Response (HAZWOPER)

In construction work the reality is that you could be exposed to chemical spills or hazardous waste. During excavation operations many construction companies have been surprised by what they have uncovered. Sometimes excavation sites can become deadly.

Under HAZWOPER, your employer can either participate in the handling of emergencies involving hazardous waste or chemical spills, or immediately evacuate workers and call in trained HazMat teams. Under no circumstances does OSHA permit personnel to respond to a chemical spill or hazardous waste cleanup without appropriate training.

How Can HAZWOPER Involve You?

Very specific training requirements come with preparing for accidental chemical releases. OSHA has set up a formal training schedule for emergency responders under HAZWOPER regulations, with training levels ranging from awareness training for first response to technical training for those with responsibility for solving problems associated with spill cleanup.

What Should You Do in Case of a Spill?

Whether it's a solid or a liquid spill, remember that you can be exposed to toxic dust or vapor without even knowing it. If you are properly trained, act with care and speed. However, if you have not received training, **do not** respond to a chemical spill. Instead, follow your company's EAP for reporting hazardous spills and evacuating.

While it is vital to avoid panic, it is equally vital to get people out of harm's way as quickly as possible. Assess site hazards and act only when you do not endanger yourself. You want to help your fellow employees, but you won't do that if you become a victim yourself.

Some of the things you can do before help arrives are:

- Determine the potential hazards (look at the material safety data sheet),

- Know about spill equipment and safety personnel,

- Know the exits and escape routes,

- Know the location of fire extinguishers, and

- Know first aid or where to get first aid equipment.

Act Responsibly in Chemical Spill Emergencies

Safety becomes extremely important when hazardous chemicals are spilled. A mistake here can be deadly. Use the buddy system, whether you're part of the emergency response team or not. Don't ever enter a chemical emergency situation alone.

Different chemicals will require different levels of protective clothing or other precautions. Don't touch any spills without protection. Avoid the contaminated clothing of injured persons. Certainly, if the chemicals involved aren't hazardous, respond immediately within your abilities.

Perform only those emergency response tasks for which you possess adequate training. Check the area for potential hazards such as electrical cords or wires near the spill or obstacles in the path of the emergency response team. Check for injuries and notify emergency medical personnel.

Decontaminate victims, if possible. Cooperate with emergency personnel when they arrive. Pass on any information you've gathered.

Evacuation

The first indication that there is a fire or other emergency is often the sound or sight of an alarm. Employee alarm systems are designed to provide a warning for necessary emergency action, as called for in your company's emergency action plan, or for your safe escape from the worksite. Your company's plan spells out your role. Find out what you are expected to do in case of a fire or other emergency.

Your company may conduct fire drills. During these drills, learn where exits are and how to evacuate the affected area. Practice getting to your designated head count area as quickly, orderly, and safely as possible.

For most people, when they hear a fire alarm or other emergency warning sound, they make their way to the nearest exit and meet at a predesignated spot. If the exits are blocked or cluttered, exiting can be difficult or even dangerous. Therefore, keep emergency exits clear for easy access. When a fire occurs, seconds count. Know where the exits are and how to get to them safely.

Exits should be marked with a readily visible sign, with no other signs or distracting objects near it. If you cannot immediately see an exit, you should be able to see readily visible signs directing you to it. Doors, passages, or stairways that are *not* exits should be identified with a sign reading "Not an Exit" or similar designation.

Emergency Follow-up

Following an emergency, OSHA must be notified if the incident resulted in any fatalities or if three or more persons were hospitalized. If a chemical spill is significant, the National Response Center must be notified as well.

The final activity following any emergency is to review and evaluate all aspects of what happened and what may happen as a result. Because an account must be made of the incident and it must be accurate, authentic, and complete, be prepared to cooperate. The events of the incident should be recorded in chronological order with each entry signed.

Work at Working Safely

Take these precautions before an emergency:

- Determine the potential hazards in any emergency situation before acting,

- Keep work areas clean and clutter free,

- Know where to find and how to use emergency equipment and safety personnel,

- Know the location of fire extinguishers,

- Know first aid or where to get supplies (only give first aid if you are qualified),

- Know the locations of exits and escape routes, and

- Be familiar with your company's emergency action plan.

Employee _____

Instructor_____

Date _____

Company _____

EMERGENCY RESPONSE REVIEW

1. The emergency action plan must:
   a. Be kept at each worksite.
   b. Be made available to all employees.
   c. Be in writing if your company has 11 or more employees.
   d. All the above

2. Fire is made up of four basic components including fuel, heat or an ignition source, _____, and a chain reaction.
   a. Combustion.
   b. Wind.
   c. Oxygen.
   d. Carbon Monoxide.

3. Using the wrong type of extinguishing agent on a fire may increase the intensity of the fire.
   a. True.
   b. False.

4. Most fire extinguishers have about _____ seconds of operation time.
   a. 2-3 seconds.
   b. 8-10 seconds.
   c. 20-25 seconds.
   d. 45-50 seconds.

5. If you come upon a chemical spill you should:
   a. Not respond to the spill unless you have the proper training.
   b. Get out of harm's way as soon as you can.
   c. Don't touch any spills without the proper protection.
   d. All the above.

6. When you hear a fire alarm you should:
   a. Stay where you are and wait for help.
   b. Make sure it is the fire alarm and not some other type of warning (like a tornado alarm).
   c. Grab a fire extinguisher and go looking for the fire.
   d. Attempt to put the fire out by aiming the fire extinguisher at the top of the flames.

7. The best type of fire extinguisher to use on combustible metals is a class:
   a. A.
   b. B.
   c. C.
   d. D.

8. The two major jobsite fire hazards you need to be aware of are:
   a. Combustible materials and ignition sources.
   b. Flammable and combustible liquids and heating devices.
   c. Inadequate fire extinguishers and lack of training in how to use them.
   d. None of the above.

9. A properly trained person should use the _____ method of fire extinguisher use.
   a. GLASS.
   b. SASS.
   c. PASS.
   d. LASS.

10. What basic safety procedure do you need to know when an emergency occurs?
    a. How to get to the appropriate exit.
    b. Where to get first aid.
    c. Where fire extinguishers are located.
    d. All of the above.

ERGONOMICS AND BACK SAFETY: WORK THAT FITS PEOPLE

Ergonomics is the science of fitting jobs to people. It considers physical abilities and limitations as well as other human characteristics. When ergonomics is applied correctly at your worksite, you can work more comfortably, safely, and efficiently.

In this chapter, we will focus largely on lower back pain, a condition commonly caused by a lack of ergonomics.

Back Disorders

Your spine consists of 24 vertebrae, each separated from the next by soft discs that act as shock absorbers when the vertebra move. Your abdominal muscles as well as other muscles and ligaments that run along the spinal column support your spine. If your spine is not properly supported by these muscles, a quick twist or off-balanced lift can easily result in a low back injury. Because your lower back (the lumbar area) carries most of your weight, it is usually the first damaged area of your back.

Back disorders are frequently caused by:

- Repeated lifting;

- Sudden movements;

- Whole body vibration;

- Lifting and twisting at the same time;

- Bending over for long periods of time;

- Poor physical condition, stress, and aging; and

- Poor posture.

Sometimes back injuries come from lifting heavy or awkward objects one time. Lots of back injuries, however, result not from a single lift, but from relatively minor strains that occur over time. As you repeat a particular irritating movement, minor injuries begin to accumulate and weaken affected muscles or ligaments. Eventually, a more serious injury may occur.

Other Musculoskeletal Disorders (MSDs)

An MSD is an injury/disorder of the muscles, tendons, joints, spinal discs, nerves, ligaments, or cartilage. Most occur from continual wear and tear. They do not include injuries caused by slips, trips, and falls. Generally, MSDs in construction workers affect the hands, wrists, shoulders, neck, upper and lower back, and the hips and knees. Different types of MSDs are often associated with different types of construction work. Some of the most common MSDs include:

- **Sprain**—Injury or tear to a ligament. Ligaments attach one bone to another or support organs.

- **Strain**—Injury to muscles that have been stretched or used too much. Symptoms include muscle irritation, pain, and discomfort.

- **Degenerative disc**—Damage to the gel-like cushions between the bones in the spine. Released gel presses on the nerve. Symptoms include numbness, pain, and weakness, usually in the leg and hips, but sometimes in the arms and upper back.

- **Tendinitis**—Inflammation and soreness in tendons caused by repeated movement of a joint. Most common for the wrists, palm side of any finger, thumbs, and elbows. Symptoms include burning pain or dull ache, swelling or puffiness, and snapping or jerking movements. Workers affected include roofers, sheet metal workers, masons, iron workers, rodmen, and workers who use a staple gun, screw driver, or a tool too large or small for the hand.

- **Raynaud's syndrome**—Nerve and blood vessel damage in the hands caused by the use of vibrating hand tools such as power hand tools, power snips, grinding wheels, gas-powered circular and chain saws, needle guns, torque wrenches, jackhammers, impact tools, and others. Symptoms include numbness and weakness in the hands and fingers, whitening of the fingers, hands, and sometimes the forearm to the elbow after exposure to vibration or cold.

- **Carpal tunnel syndrome**—Nerve disorder in the hand and wrist caused by repeated bending of the wrist, holding tools or materials tightly, and constantly pressing the wrist against a hard object. Symptoms include numbness, tingling, burning, and pain. In severe cases, there may be wasting of the muscles at the base of the thumb; a dry shiny palm; or clumsiness of the hand. Workers affected include carpenters, electricians, and sheet metal workers.

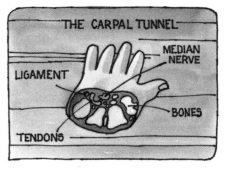

- **Thoracic outlet syndrome**—Reduced blood flow in the shoulder and arm caused by overhead work or carrying heavy items in the hands with the arms straight down. Workers affected include scaffold erectors, insulators, and painters.

- **Carpet layers' knee**—Repeated use of knee kickers while laying carpeting. Symptoms include knee pain and sprains. Workers affected include floor layers, tile setters, roofers, electricians, sheet metal workers, carpenters, and insulators.

Ergonomic Risk Factors

Musculoskeletal disorders develop as a result of repeated exposure to ergonomic risk factors including:

- **Force**—Amount of physical effort to do the task. Examples include lifting and pushing.

- **Repetition**—Doing a series of motions over and over. Examples include screw driving or nailing.

- **Awkward postures**—Positions of the body that require muscle activity to maintain. Examples include overhead work, twisting, and squatting.

- **Static postures**—Physical exertion to hold a position throughout the task. Examples include gripping tools, holding arms out or up, and prolonged standing.

- **Vibration**—Jolting motion of the body. Examples include using an electric drill or wood planer, driving a truck, and operating a jackhammer.

- **Contact stress**—Occasional, repeated, or continuous contact between sensitive body tissue (usually the fingers, palms, forearms, thighs, shins, and feet) and a hard or sharp object. Examples include pressing on tool handles, hand hammering, and sitting without knee room.

- **Cold temperatures**—Exposure to excessive cold while working. This reduces dexterity and hand sensitivity. Examples include gripping cold tools.

If a given task involves several of these risk factors, it is more likely to result in the onset of a musculoskeletal disorder. The goal then is to eliminate or reduce some, if not all, of the risk factors at your jobsite.

Preventing Back Disorders and Other MSDs

If your company discovers possible ergonomic risk factors at a site, it has several ways to eliminate or reduce them.

Engineering Controls

The best way to fix ergonomic problems is to implement engineering controls to make the job fit you, not force you to fit the job. These controls involve designing or redesigning work areas, tools, or equipment to fit you and could include:

- Adjusting work surface heights;

- Making work more accessible, such as bringing overhead work down or raising yourself to the work level using a platform, scaffold, or other means;

- Changing work area layouts, such as moving vehicle mirrors and seats to avoid awkward positions;

- Providing foot rests;

- Reducing sizes or weights of objects lifted, such as breaking loads into smaller, lighter ones;

- Installing mechanical aids, such as using tools to reach far away objects;

- Padding hard or sharp surfaces, such as tools, materials, and seats;

- Providing carts; and

- Designing for women.

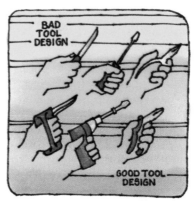

Administrative Controls

Administrative controls include reducing the duration, frequency, and severity of exposure to jobsite hazards. These include:

- Job rotation,

- Shortened shift lengths,

- Limiting overtime,

- Adding more frequent breaks,

- Alternating repetitive tasks with less repetitive ones,

- Decreasing production rates, and

- Increasing the number of workers assigned to a task.

Work Practice Controls

Even with good engineering and administrative controls in place, it is important that you follow good work practices such as maintaining proper posture, using good lifting and other techniques, having a balanced diet, performing strengthening exercises, and understanding the limitations of support devices.

Proper Posture

To put the least amount of stress and strain on your muscles and bones, use these posture techniques:

- **Standing**—Tuck in your chin and relax your shoulders. Keep feet at least a foot apart. Stand with one foot on a small stool and switch feet. Move about whenever possible.

- **Sitting**—Keep your head directly over your shoulders. Relax your shoulders. Ensure the chair back is supporting your lower back. Keep your knees at the same level as your hips or slightly lower. Sit as close to your work as you can.

Proper Lifting

Many low back strains and injuries can be avoided by this basic procedure for good lifting:

1. **Size up the load before trying to lift it.** Test the weight by lifting at one of the corners. Get help or use a device if the load is too heavy.

2. **Make sure the load can be carried to its destination** before attempting the lift. Make sure the path is clear and safe.

3. **Bend the knees.** Place your feet apart and close to the object. Center yourself over the load, then bend your knees and get a good hand hold. Lift straight up, smoothly. Do not bend at the waist. Allow your legs to do the work.

4. **Do not twist or turn the body** once a lift is made. If you must turn the body, do so by changing foot positions. Keep the load steady and close to the body. Do not carry a load above your head or on your side.

5. **Set the load down properly.** Bend the knees, keeping your back upright, letting your legs do most of the work. Do not let go of the load until it is on the floor.

6. **Always push an object, rather than pull it.** Pushing puts less strain on the back and is safer should the object tip.

Proper Techniques

To prevent MSDs affecting the neck, arms, wrists, and hands, try these techniques:

- Use kneepads or cushions for kneeling jobs;
- Use the full-hand grip instead of the pinch grip; and
- Maintain tools and equipment, i.e., sharpen blades.

Proper Diet

A proper diet can prevent injury:

- Drink 8 glasses of water a day to reduce tearing injuries and prevent stiffness.

- Eat a well-balanced diet for energy. Injuries occur when you are mentally and physically tired.

- If you have pain, decrease caffeine intake. Caffeine increases muscle sensitivity to pain.

Strengthening Exercise

Stretching can make your back stronger, more flexible, and more resistant to injury. Work on your back, thigh, buttock, and hamstring muscles. Bend and stretch these muscles, holding them for at least 15 seconds, without bouncing.

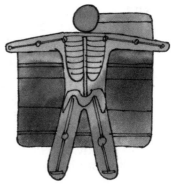

Training

Your company should provide ergonomics and back safety training. You should learn:

- About your job's ergonomic hazards;

- Ways to protect yourself including how to properly use equipment, tools, and machine controls, as well as the correct way to do a variety of job tasks (this includes lessons on proper posture and lifting);

- To recognize when you have signs and symptoms of MSDs;

- How to report MSD signs and symptoms so that problems can be identified at an early stage when treatment is likely to be more successful; and

- Engineering and administrative controls the company has implemented.

Work at Working Safely

Following simple ergonomic principles helps reduce your risk of injury to your musculoskeletal system. If you have ever had back pain you know how important this is. If you have not suffered from back pain or other musculoskeletal disorders, following these principles will help assure that you never will.

Employee _____

Instructor_____

Date _____

Company _____

ERGONOMICS AND BACK SAFETY REVIEW

1. The discipline of ergonomics is:
    a. Arrangement of the job to fit the person.
    b. Job safety analysis.
    c. Hazard analysis.
    d. Both a. and b.

2. Trigger finger, tendinitis, and De Quervain's disease are:
    a. Carpal tunnel syndrome.
    b. Raynaud's syndrome.
    c. Musculoskeletal disorders.
    d. All of the above.

3. A disorder that affects the nerves of the hands and wrists is:
    a. Tendinitis.
    b. Trigger finger.
    c. Raynaud's syndrome.
    d. Carpal tunnel syndrome.

4. Tendonitis is an inflammation of the tendon in which the tendon is:
    a. Repeatedly tensed from overexertion.
    b. Tensed from vibration.
    c. Unaccustomed to a particular use.
    d. All of the above.

5. Raynaud's syndrome is a condition where the blood vessels of the hand are damaged due to repeated exposure to long periods of vibrations.
    a. True
    b. False

6. Ergonomically designing work stations, tools, and equipment so the job fits you is a(n):
   a. Engineering control.
   b. Work practice control.
   c. Administrative control.
   d. Medical management technique.

7. An example of an engineering control that reduces the force an employee uses to lift an object is:
   a. Allowing the worker to take rest breaks.
   b. Telling the worker to lift with his legs, not his back.
   c. Providing a mechanical lifting device.
   d. Assigning more workers to the job.

8. All of the following are workplace risk factors except:
   a. Awkward position.
   b. Repetitive motions.
   c. Nonvibrating, ergonomically designed tools.
   d. Twisting motions.

9. Following proper lifting procedures is a(n):
   a. Work practice control.
   b. Administrative control.
   c. Engineering control.
   d. None of the above.

10. To prevent injuries to the back:
    a. Do regular and moderate physical exercise.
    b. Excessively twist, bend, and reach.
    c. Sit for prolonged periods.
    d. Be in poor physical condition and have poor posture.

EXCAVATIONS: NOW DIG THIS

Working around and in excavations is one of the most dangerous jobs in the construction
industry. There are many hazards
you can run into but most can be
placed into three categories:
underground utilities, confined
space hazards, and cave-ins. In
2002 there were 106 injuries from
cave-ins and 34 fatalities.

Cave-ins are usually the result of
unsafe work habits, changes in weather that change soil stability, vibrations caused by construction activity, and movement of
buildings near excavation sites. You must always be alert for
changing conditions!

What Is an Excavation?

An excavation is any man-made cut, cavity, trench, or depres-
sion in the earth's surface formed by earth removal. This can
include excavations for anything from a sewer line installation
to multi-lane interstate highways.

Where Are the Regulations?

Specific regulations for excavation work, as required by OSHA,
are found in 29 CFR 1926.650-.652.

Before Digging Begins

Before you start to excavate, a member of management or a
"competent person" will need to:

* Contact utility companies and the property owner to ensure
  underground installations are found.

- Make sure underground installations are protected, supported, or removed as necessary to safeguard employees.

- Remove or secure any surface obstacles, such as trees, rocks, and sidewalks, that may create a hazard.

- Classify the type of soil and rock deposits at the site. One visual and at least one manual analysis must be made.

The Competent Person

A competent person is a company employee who:

- Is trained in and capable of identifying existing and predictable hazards which are unsanitary, hazardous, or dangerous.

- Is responsible for performing the soil classification analysis.

- Has the authority to take prompt corrective measure to eliminate any hazards.

- May be responsible for coordination and direction of emergency response.

- Must inspect the excavation and adjacent areas at least once a day for possible cave-ins, failures of protective systems and equipment, hazardous atmospheres, or other hazardous conditions.

Soil Classification

Before you can work in an excavation, the soil type must be determined. The soil must be classified as: stable rock, type A, type B, or type C soil. It is not uncommon to find a combination of soil types at an excavation site. Soil classification is used to determine the need for a protective system.

The definitions of the different soil types are:

Stable rock—Natural solid mineral material that can be excavated with vertical sides and will remain intact while exposed.

Type A soil—Examples include clay, silty clay, sandy clay, clay loam, and sometimes silty clay loam and sandy clay loam.

Type B soil—Examples include silt, silt loam, sandy loam and sometime silty clay loam and sandy clay loam.

Type C soil—Examples include granular soils like gravel, sand, loamy sand, submerged soil, and soil from which water is freely seeping, and submerged rock that is not stable.

Why Are Protective Support Systems Important?

You must be protected from cave-ins by a protective system designed according to OSHA standards. There are many factors involved in designing a protective system including soil classification, depth of cut, water content of soil, changes due to weather and climate, and other operations in the vicinity.

Types of protective systems include:

- Proper sloping and/or benching of the sides of the excavation.

- Supporting the sides of the excavation with timber shoring or aluminum hydraulic shoring.

- Placing a shield between the sides of the excavation and your work area.

Supervisors can choose the most practical design for the work being done. Once a system is chosen, the required performance criteria must be met for that system.

The Four-Foot and the Five-Foot Rule

Sometimes rules can be confusing. One of the confusing points in excavations is the four-foot and the five-foot rule.

The four-foot rule refers to your means of escape and says: An exit must be provided if the excavation is four feet deep or greater. This exit must be within 25 feet of every worker.

The five-foot rule refers to when you do not need a protective system and says: A protective system isn't needed if the excavation is less than five feet deep (provided a competent person determines there is no indication of a potential cave-in). A protective system is also not needed if the excavation is made entirely in stable rock.

Emergency Response During an Excavation

Emergency rescue equipment is required by OSHA when a hazardous atmosphere exists or may reasonably be expected to develop during excavation work (see the chapter on confined spaces for more information). However, your company should have emergency response procedures in place, and rescue equipment ready, in case any accident occurs. The procedure should include:

- Who will provide immediate jobsite rescue and aid?

- Who will notify the authorities and rescue personnel?

- Who will meet, advise, and direct rescue personnel?

- What emergency response equipment will be available on the jobsite, where will it be kept, and who is trained to use it?

If there is an accident at the excavation site, the time between when the accident happened and when rescue personnel arrive is critical. Any assistance that you can give the victim, with out endangering yourself, should be done immediately. Avoid using heavy equipment in a rescue attempt of a trapped person. You must use extreme caution in this situation.

Work at Working Safely

The following safety rules are ones that you should know and practice while working in trenches. Excavations can be deadly, but they do not have to be. Be conscientious at all times.

- Know your job responsibilities. Ask questions.

- Always wear the proper safety equipment required.

- Know your company's emergency response procedures.

- In excavations greater than four feet, and where hazardous atmospheres exist, or could exist, be sure your competent person tested the air before you enter that excavation.

- Keep materials or equipment that might fall or roll into an excavation at least two feet from the edge.

- Wear a warning vest or other suitable clothing, marked with or made of reflectorized or high-visibility material, when you are exposed to vehicle traffic.

- Use warning barricades, hand or mechanical signals, or stop logs, to alert equipment operators of the edge of an excavation.

- Be sure you have adequate protection from falling rock, soil, or other materials and equipment.

- Do not work in excavations where water has accumulated, or is accumulating, unless adequate precautions have been taken.

- Do not cross over an excavation unless walkways are provided. Guardrails must be provided if the walkway is six feet or more above the bottom of the excavation.

NOTES

Employee _____

Instructor_____

Date _____

Company _____

EXCAVATIONS REVIEW

1. Excavation employees must wear high-visibility warning vests when they:
   a. Work under loads being handled by digging equipment.
   b. Signal directions to equipment operators.
   c. Are exposed to public vehicular traffic.
   d. Work in the rain.

2. Before starting an excavation, the employer must:
   a. Contact the utility companies to inform them of the proposed work.
   b. Wait until the utility companies have exposed any underground lines.
   c. Find the exact location of underground installations.
   d. None of the above.

3. The competent person must inspect all excavations each day for:
   a. Signs that indicate a cave-in is possible.
   b. Failure of protective systems.
   c. Evidence that a hazardous atmosphere could develop.
   d. All of the above.

4. The primary hazard in excavation work is:
   a. Falls.
   b. Cave-ins.
   c. Drowning.
   d. Electrocution.

5. Make sure underground installations are protected, supported, or removed as necessary to safeguard employees.
   a. True
   b. False

6. Materials must be kept at least _____ feet from the edge of the excavation.
   a. Six
   b. Three
   c. Two
   d. Twelve

7. Soil classification, depth of cut, water content of soil, changes due to weather and other operations in the vicinity are all factors to be considered when:
   a. Continuing the excavation.
   b. Designing a protective system to protect workers from cave-ins.
   c. Deciding what equipment to use.
   d. None of the above.

8. Shoring systems support the _____ of an excavation.
   a. Top.
   b. Middle.
   c. End.
   d. Sides.

9. A competent person must perform an inspection If the excavation after every rainstorm.
   a. True
   b. False

10. Trenches that are 4 feet deep need a safe means of egress:
    a. At both ends of the trench.
    b. Not more than 10 feet from the workers.
    c. Not more than 25 feet from the workers.
    d. At one end of the trench.

FALL PROTECTION: THE BENEFITS ARE UPLIFTING

In 2002 there were a total of 272,988 injuries from falls in construction, with 714 fatalities. Most falls were from or through roofs. Falls from ladders and stairs totaled 44,593 with 137 fatalities.

Your employer has a duty to provide you with fall protection equipment and training when required. It is your duty to use the equipment and training provided.

Where Are the Regulations?

The regulations for protection against falls can be found throughout 29 CFR 1926:

29 CFR 1926	Name
.450-.454	Scaffolding (see the Scaffolds chapter of this handbook)
.500-.503	Fall protection
.1050-.1060	Stairways and ladders (see the Ladders & Stairways chapter of this handbook)

Fall Protection

The fall protection standard has three elements:

- Situations at your worksite that require protection from falling and from falling objects **(1926.501)**.

- Different types of fall protection equipment and systems your employer can use to provide you protection **(1926.502)**.

- Training requirements **(1926.503)**.

General Requirements

Your employer is required to:

- Provide equipment and training to protect you from falling off, onto, or through working levels that are six feet or more above lower levels and to protect you from falling objects.

- Ensure that when your working area is elevated, it has the required strength and structural integrity to support you and your fellow workers.

- Make sure that body belts are not used as personal fall arrest equipment. However, they can be used as positioning devices. Also, only locking type snaphooks can be used in personal fall arrest equipment.

Equipment and Systems

Your employer must provide and install all required fall and falling object protection before you begin work. The three most common methods of providing fall protection are guardrails, safety nets, and personal fall arrest systems. These are referred to as primary systems.

Guardrails

A guardrail is a barrier put up to prevent falls to a lower level. Guardrails can be used to protect you from falls from unprotected sides and edges, during leading edge construction work, through holes including skylights, from ramps, runways, or other walkways, into excavations that are not visible because of weed growth or other visual barriers, and into or onto dangerous equipment.

Some guardrail requirements are:

- The top of the top rail must be 39 to 45 inches high.

- Guardrails must be able to withstand, without failure, a 200 pound force applied within two inches of the top edge in any outward or downward direction at any point along the top edge. The top edge must not bend to a height less than 39 inches above the floor.

- Midrails must be capable of withstanding 150 pounds of inward or downward force.

- Steel or plastic banding cannot be used as top or midrails.

- If wire rope is used for top rails, it must be flagged at not more than six foot intervals with high-visibility material.

- Manila, plastic, or synthetic rope used for top or midrails must be inspected frequently for strength.

- When guardrails are used around holes, all unprotected sides and edges must be protected.

Safety Nets

Safety nets are used as protection at unprotected sides, leading edges, working on the face of formwork or reinforcing steel, overhead or below surface bricklaying, work on roofs, precast concrete work, residential construction, and wall openings.

Some of the requirements for using safety nets are:

- Installed nets as close as possible under where you are working, but never more than 30 feet away.

- Extend nets outward from the edge of your work area:

Vertical drop from your working surface	Minimum distance from the outer edge of the working area
Up to 5 feet	8 feet
More than 5 up to 10 feet	10 feet
More than 10 feet	13 feet

- After your safety net is installed, drop test or certify it before use. Also drop test or certify a net whenever it is relocated, repaired, and every six months if it is in one location that long.

- Inspect nets at least once a week for wear, damage, and other deterioration. Replace defective nets or components before use.

- Remove materials, scrap pieces, equipment, and tools that have fallen into the net as soon as possible and at least before the next work shift.

Personal Fall Arrest Equipment (Harnesses)

Personal fall arrest equipment arrests a fall, but doesn't prevent a fall, when you are working around unprotected sides and edges, leading edge work, in hoist areas when loading or unloading materials, form and reinforcing steel work, overhead or below surface bricklaying, work on low-sloped or steep roofs, precast concrete work, residential construction, and wall openings.

In case you do fall, the fall protection equipment must be rigged to limit your fall: to a free fall of not more than 6 feet, to a deceleration distance of no more than 3.5 feet, and to prevent you from contacting any lower level.

FALL PROTECTION

Some other requirements for personal fall arrest equipment are:

- A horizontal lifeline must be designed, installed, and used under the supervision of a qualified person.

- Lanyards and vertical lifelines must have a minimum breaking strength of 5,000 pounds.

- Only one worker can be attached to a vertical lifeline.

- Protect your lifeline from, and inspect your lifeline for, cuts or abrasions before and during work.

- Ropes and straps (webbing) used in body harnesses must be made from synthetic fibers.

- For a full body harness the attachment point must be located in the center of your back near your shoulder or above your head.

- Your employer must have a rescue plan and be able to rescue you promptly, or ensure you are capable of rescuing yourself.

- Inspect your fall arrest equipment prior to each use for wear, damage, and other deterioration. Do not use defective components.

- Do not attach your personal fall arrest equipment to:

  - Anchorages being used to support or suspend platforms.

  - Guardrails.

  - Hoists, except as specified in the rules.

Other Fall Protection Systems

The fall protection rule lists other systems and equipment your employer can use in certain situations. Some of them are:

- **Safety monitoring system**—Used when working on low-slope roofs only. It must be used with a warning line system. The only exception is that a safety monitoring system can be used alone when the roof is 50 feet or less in width, as the OSHA rule determines width.

- **Covers**—Required for holes, including skylights.

- **Warning lines**—Must be used with another protective system such as guardrails, safety nets, personal fall arrest equipment, or safety monitoring procedures.

- **Positioning devices**— Used on the face of formwork or reinforcing steel structures and other situations where hands must be free to work.

Protection from Falling Objects

Falling objects are a major hazard around construction sites. You should wear your hard hat at all times when falling objects are a possibility.

Also protect your coworkers by one or a combination of the following methods:

- Toeboards along the edge of an overhead walking/working surface.

- Guardrails that have all openings small enough to prevent passage of potential falling objects.

- Proper storage methods during bricklaying, roofing, and related work.

- Canopies and barricades.

Steel Erection

OSHA's steel erection standard has requirements for protecting workers from falls. Once iron workers are 15 feet above a lower level they must use adequate fall protection equipment. However, there are two exceptions to this rule: Connectors working at heights between 15 and 30 feet, and employees working in a controlled decking zone between 15 and 30 feet, do not need fall protection equipment when special provisions in the OSHA regulations are followed.

Training Requirements

You must be trained by a competent person, anytime you could be exposed to fall hazards. Training includes:

- Recognizing fall hazards at your worksite and how to minimize them.

- Correct procedures for erecting, maintaining, disassembling, and inspecting the fall protection equipment and systems you will use.

- Proper use and operation of the fall protection systems.

- Your role in a safety monitoring system, if it is used.

- Limitations of the use of mechanical equipment when working on low-sloped roofs.

- An understanding of the OSHA fall protection rules.

Your employer must:

- Prepare a written certification of your training.

- Retrain you when you don't understand something, when you move to a different workplace, or when new equipment is introduced.

Work at Working Safely

Contrary to popular belief, it is possible to prevent injury and death due to falls. No one should consider accidents to be a part of the cost of doing business. Rework and repair of people is much more difficult than it is with products and equipment. Therefore, use fall protection equipment correctly, protect it from jobsite hazards, and inspect it prior to each use. Doing this may save your life.

Employee _____

Instructor_____

Date _____

Company _____

FALL PROTECTION REVIEW

1. Under OSHA's fall protection rule for construction, workers must be protected from falls when they are on walking/working surfaces that are _____ or more above a lower level.
    a. 4 feet.
    b. 6 feet.
    c. 10 feet.
    d. 24 feet

2. Conventional fall protection systems include:
    a. Guardrail systems, controlled access zones, and scaffolds.
    b. Safety net systems, fall protection plans, and positioning devices.
    c. Personal fall arrest systems, warning line systems, and the buddy system.
    d. Guardrail systems, safety net systems, and personal fall arrest systems.

3. In a guardrail system, when wire rope is used as the top rail:
    a. It must have a diameter of at least 1/2 inch.
    b. It must be at least 47 inches above the walking/working surface.
    c. It requires high-visibility flags every 6 feet.
    d. Wire rope cannot be used as a top rail.

4. If materials, scrap pieces, equipment, or tools fall into a safety net:
    a. They must be removed immediately by the person who dropped the material.
    b. They must be removed as soon as possible—and at least before the next shift.
    c. They must remain in the net until the next weekly inspection.
    d. They will probably fall through the net eventually.

5. A personal fall arrest system must limit an employee's fall:
   a. To a free fall of not more than 6 feet.
   b. To a deceleration distance of no more than 3.5 feet.
   c. So the employee does not contact any lower level.
   d. All of the above.

6. A positioning device system:
   a. Can not be used.
   b. Can be used to hoist materials.
   c. Allows the worker to be supported on a wall to work with both hands free.
   d. Cannot use a body belt.

7. In order to use any fall protection system, you must be trained by your company's competent person.
   a. True
   b. False

8. All walking and working surfaces:
   a. Must have guardrails.
   b. Need toeboards and guardrails.
   c. Must be strong enough to support the workers.
   d. Must be kept clear of stored materials.

9. Which of the following is **not** included in OSHA's steel erection standard for protecting iron workers?
   a. Requiring workers to use adequate fall protection at heights of 15 feet above a lower level.
   b. Allowing connectors working at heights between 15 and 30 feet to not use fall protection if special provisions are met.
   c. Requiring workers to use adequate fall protection at heights of 10 feet above a lower level.
   d. Allowing employees working in a controlled decking zone between 15 and 30 feet to not use fall protection if special provisions are met.

10. Safety nets must be drop tested every time they are relocated.
    a. True
    b. False

FIRST AID & BLOODBORNE PATHOGENS: PROTECT YOURSELF WHILE GIVING AID

Emergencies can happen, anywhere, anytime to anyone. For this reason, it is important to have someone who knows what to do in an emergency. That person should know first aid.

> **First aid:** Emergency treatment of injury or sudden illness before professional medical care is available.

Where Are the Regulations?

The OSHA regulation for first aid is found in 29 CFR 1926.50, Medical Services and First Aid. Your company has an obligation to provide first aid equipment, training, and personnel when a hospital or clinic is not close enough to provide help within three to four minutes. That's because life-threatening injuries such as stoppage of breathing and severe bleeding can kill someone within that time period.

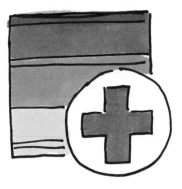

First Aid Provider

It is important to have someone who knows what to do in an emergency. It is just as important to know what not to do. Every trained person must know his or her responsibilities and limitations. In other words, if you are not qualified to help someone who is injured, you must get someone who is.

First aid providers are people who are occupationally required to be trained in first aid even though they may not be specifically obligated by law to perform first aid and are protected by the "Good Samaritan" laws. A first aid provider uses a limited amount of equipment to assess a victim and provide life support and care while awaiting arrival of emergency services.

Good Samaritans include all those who voluntarily provide assistance in an emergency. The assistance a Good Samaritan provides may be something such as performing CPR, tying a bandage around someone's leg or arm to stop the bleeding, or any other emergency care or treatment.

First Aid Response

If you are the first person to reach an injured or ill person, here are some of the basics of immediate first aid:

- **Call for help**—If you are not alone, have someone go for help. If you are alone or not qualified to give first aid, sometimes the immediate priority is leaving the victim to get help.

- **Analyze the situation**—Get the victim out of danger without putting yourself in danger.

- **Do not move the victim**—There may be a chance of neck or spine injury. If it is necessary to move the victim in a life-threatening situation, do so carefully.

- **Look for signs of life**—If you are trained, check the ABCs—clear the *Airway*, check for *Breathing* and perform rescue breathing, and then check for *Circulation* and perform CPR if necessary.

- **Control heavy bleeding by applying pressure**—Do not apply a tourniquet unless the victim is in danger of bleeding to death and you have been trained to do so.

- **Treat for shock**—Signs of shock include cold, pale skin, a rapid, faint pulse, nausea, rapid breathing and weakness.

To treat for shock, keep the victim lying down, cover him/her only enough to maintain body heat, don't move the victim unless absolutely necessary, and get medical help immediately.

- **Treat for burns**—For small burns, gently soak the burn in cold water or pour cold water on the burn. Do not treat large burns with water unless they are chemical burns. Cover the burn with a dry, sterile bandage. Provide artificial respiration as needed. Seek medical attention. Some chemicals should not be flushed with water, but neutralized by other means—see the chemical label.

- **Treat for chemical in the eye**—Quickly flush the eyes with lots of water for at least 15 minutes (for best results, do so at an eyewash station, emergency shower, or hose). Try to force the eyes open to wash the chemical out. Do not bandage eyes. Seek medical attention.

- **Treat for fracture**—Do not move the victim unless you absolutely have to. This is especially important if you suspect a neck or back injury. Get medical help.

Note: These steps are not a replacement for formal training in first aid or CPR.

First Aid Kits

Your company must provide an easily accessible, weatherproof first aid kit. This first aid kit must be checked before going out to the jobsite and at least weekly to ensure supplies are replaced as used.

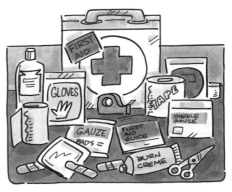

Most kits will contain some basic items to treat a number of injuries quickly and easily, including:

Bandages, gauze, and compresses	Eye dressing	Tweezers
Adhesive tape.	Eye wash solution	Latex or similar gloves
	Ammonia inhalants	Instant cold packs
Metal and aluminum splints	Antiseptic swabs	Blanket
Tourniquet	Burn treatments	Instructions for giving first aid
	Scissors	

In addition, the telephone numbers of your company physician, hospital, and ambulance service must be posted in plain view. Your company must provide transportation of an injured person to a physician or hospital if an ambulance service is not available.

Accident Reporting

After the immediate needs of a workplace accident, injury or illness emergency have been dealt with, you should report the event to a supervisor or manager. Extremely minor injuries, like a small bruise, may not need to be reported. But report to a supervisor, manager, or other designated person at your company any workplace accident, injury or illness involving:

• Professional treatment,

• Time away from work, or

• A near miss of a more serious accident.

Report these even if the injury or illness does not become apparent until after the cause (e.g., back pain may take a long time to develop).

Find out if your company has specific procedures and/or forms for accident reporting and investigation; it's a common thing for companies to have. If they do have a specific policy or form, use it for reporting or investigating accidents.

Bloodborne Pathogens

As a first aid provider, you must be aware of the risks associated with giving assistance in a medical emergency. One risk is exposure to infectious diseases through blood and other body fluids. These infectious diseases include the Hepatitis B Virus (HBV) and the Human Immunodeficiency Virus (HIV). They are also called bloodborne pathogens.

The Bloodborne Pathogens Standard (29 CFR 1910.1030) limits "occupational exposure" to blood and other body fluids. Occupational exposure is reasonably anticipated skin, eye, or nasal membrane contact with blood or other body fluids that may result from one's job.

When workers such as first aid providers are occupationally exposed to blood or other potentially infectious material, the standard requires that companies develop an exposure control plan. This plan explains how the company will protect workers from bloodborne pathogens. If you are occupationally exposed, ask your supervisor for a copy of the exposure control plan (if it has one), read it, and be sure you understand it.

Note: Although the Bloodborne Pathogens Standard does not apply directly to the construction industry, OSHA is applying the General Duty Clause of the Occupational Safety and Health Act when jobsite employers don't provide safe working environments with respect to bloodborne pathogens.

Engineering Controls

Engineering controls eliminate or minimize exposure hazards at their source. These include resuscitation bags and facilities for hand washing.

Good Work Practices

Take these steps to prevent infection:

- Wash hands immediately after removing gloves or other protective equipment, and after any hand contact with blood or potentially infectious fluids.

- Do not eat, drink, apply cosmetics, or handle contact lenses in areas where there is a potential for exposure.

- Avoid spraying or splashing of blood or body fluids; and

- Assume that all human blood and human body fluids are infectious. Many people who carry infectious diseases have no symptoms and may be unaware that they have a problem. This method of infection control is called "universal precaution."

Personal Protective Equipment

 Personal protective equipment (PPE) is the final defense against any unexpected hazard. This specialized clothing and equipment, provided by your company free of charge, may include single-use gloves, face shields or masks, eye protection, and pocket masks. Wearing PPE can greatly reduce potential exposure to all bloodborne pathogens.

Hepatitis B Vaccine

The greatest bloodborne risk is infection by HBV. Your company must make the hepatitis B vaccine available to you if you have an occupational exposure risk. If you don't wish to be vaccinated, you must sign a declination form. If you change your mind at a later date, you must still be provided with the vaccine.

If You Are Exposed . . .

If you are exposed to blood or other body fluids, report the exposure to your company so that it can evaluate the circumstances surrounding the incident and arrange immediate and confidential treatment. How the exposure occurred will be documented. Your blood will  be tested with your consent, and the source individual will be tested, if possible. Any reported illness will be evaluated, and counseling will be provided.

You should take the following steps to reduce the possibility of exposure:

- Clean and decontaminate equipment and work areas as soon as possible after contact with any blood or potentially infectious fluids;

- Remove and replace protective coverings when contaminated or at the end of each shift if there is a possibility of contamination during the shift; and

- Handle contaminated clothing as little as possible. Clothing must be bagged where it is contaminated. Wet clothing must be placed in leakproof bags. Wear gloves if you handle contaminated clothing.

Work at Working Safely

In emergency situations, prompt, properly given first aid can mean the difference between life and death, rapid versus prolonged recovery, or temporary versus permanent disability. Know where first aid kits and emergency eyewash stations are before an incident occurs, and understand and follow the universal precautions when dealing with blood.

NOTES

Employee _____

Instructor_____

Date _____

Company _____

FIRST AID & BLOODBORNE PATHOGENS REVIEW

1. Your company must provide first aid equipment, training, and personnel when a hospital is not close enough to provide help within _____ minutes.
   a. 10.
   b. 15.
   c. 3-4.
   d. 8.

2. First aid providers are protected by the _____ law.
   a. OSHA General Duty.
   b. Good Samaritan.
   c. First Responder.
   d. Equal Opportunity.

3. The first thing you should do when you reach an injured or ill person is:
   a. If you are not alone, have someone else go to get help.
   b. If you are alone, you must never leave the victim to get help, even if you are not qualified to provide the needed first aid.
   c. If you are not qualified to give the needed first aid, you have to leave the victim to get help.
   d. Both a and c.

4. It is OK to apply a tourniquet instead of simply applying pressure.
   a. True.
   b. False.

5. You should only move an injured person if:
   a. They have a broken leg.
   b. They are in shock.
   c. They are exposed to a life-threatening situation.
   d. None of the above.

6. The correct amount of time to flush the eyes after being exposed to chemical splash is:
   a. 10 minutes.
   b. 15 minutes.
   c. 5 minutes.
   d. None of the above.

7. Training for CPR is not necessary.
   a. True
   b. False

8. Personal protective equipment is the _____ defense against any unexpected hazard.
   a. first.
   b. last.
   c. best.
   d. only

9. Assume that all human blood and human body fluids are infectious.
   a. True
   b. False

10. Engineering controls eliminate or minimize exposure hazards at their source.
    a. True
    b. False

HAZARD COMMUNICATION: THE RIGHT TO KNOW LAW

For one out of every four workers, working with hazardous chemicals is a daily experience. In many cases, the chemicals you deal with may be no more dangerous than those you use at home. But in the workplace, exposure is likely to be greater, concentrations higher, and exposure time longer. Thus, potential danger is greater on the job.

Where Are the Regulations?

OSHA implemented the Hazard Communication Standard to help control chemical exposure on the job. The regulation is more commonly known as "HazCom" or the "Right to Know Law," and is found at 29 CFR 1926.59.

The HazCom rule says that you have a right to know what chemicals you are, or may be, exposed to. It is important that you are aware of the standard and how it protects you. The standard requires that all chemicals at your worksite be fully evaluated for possible physical or heath hazards. And, it mandates that all information relating to these hazards be made available to you.

The areas specifically covered in the standard include:

- Determining the hazards of chemicals,

- Material safety data sheets (MSDSs),

- Labels and labeling,

- A written hazard communication program,

- Employee information and training, and

- Trade secrets.

The hazard communication standard is intended to cover all employees who may be exposed to hazardous chemicals under normal working conditions or where chemical emergencies could occur. As mentioned previously, the standard applies to those chemicals which pose either a physical or health hazard.

Physical and Health Hazards

Physical hazards are exhibited by certain chemicals due to their physical properties—flammability, reactivity, etc. These chemicals fall into the following classes:

- Flammable liquids or solids.

- Combustible liquids.

- Compressed gases.

- Explosives.

- Organic peroxides.

- Oxidizers.

- Pyrophoric materials.

- Unstable materials.

- Water-reactive materials.

A *health hazard* is a chemical that may cause acute or chronic health effects after exposure. It can be an obvious effect, such as immediate death following inhalation of cyanide. But a health hazard may not necessarily cause immediate, obvious harm or make you sick right away. In fact, you may not see, feel, or smell the danger.

An acute health effect usually occurs rapidly, following a brief exposure. A chronic health effect is long, continuous, and follows repeated long-term exposure.

Some examples of chemicals which exhibit health hazards are:

Type of chemical	Example of type
Carcinogens (cancer-causers)	Formaldehyde or benzene
Toxic agents	Lawn and garden insecticides, arsenic compounds
Reproductive toxins	Thalidomide or nitrous oxide
Irritants	Bleaches or ammonia
Corrosives	Battery acid or caustic sodas
Sensitizers	Creosote or epoxy resins
Organ-specific agents act on specific organs or parts of the body	Sulfuric acid (affects skin), or asbestos (affects lungs)

The hazard communication standard doesn't apply to hazardous waste regulated by the Environmental Protection Agency, tobacco products, many wood or wood products, food, cosmetics, and certain drugs.

Material Safety Data Sheets

A material safety data sheet (MSDS) is a fact sheet for a chemical posing a physical or health hazard at your worksite. MSDSs must be in English and contain the:

- Identity of the chemical (as used on the label),

- Physical hazards,

- Health hazards,

- Primary routes of entry,

- Whether it is a carcinogen,

- Precautions for safe handling and use,

- Emergency and first aid procedures,

- Date of preparation of latest revision, and

- Name, address, and telephone number of manufacturer, importer, or other responsible party.

If relevant information in one of the categories was unavailable at the time of preparation, the MSDS must indicate that no information was found. Blank spaces are not permitted. If you find a blank space on an MSDS, contact your supervisor.

Your company must have an MSDS for each hazardous chemical it uses. Copies must be made readily available at your worksite. When you must travel between worksites during the day, MSDSs may be kept at a central location.

If there are workers from other companies at your worksite, they must be made aware of the chemicals you use and the location of your MSDSs. They must do the same for you. All MSDSs can be at a central location and managed by the general contractor.

Labels and Labeling

Containers of hazardous chemicals must be labeled in English. Information may also be presented in other languages for non-English speaking employees, but English is required. It is required that labels contain the following information:

- Identity of the hazardous chemical.

- Appropriate hazard warnings.

- Name and address of the chemical manufacturer, importer, or other responsible party.

On individual stationary containers you may use signs, placards, batch tickets, or printed operating procedures in place of labels.

When a chemical is transferred from a labeled container to a portable container, and the chemical is intended only for the use of the employee making the transfer during his or her workshift, the company is not required to label the portable container. If, however, that container is transferred to another employee, for use on another workshift, it must be labeled.

HMIS® Labeling

HMIS® stands for Hazardous Materials Identification System. It's a complete labeling program that helps your employer comply with OSHA's HazCom Standard.

The HMIS® program uses a numerical hazard rating system (the higher the number, the more dangerous the substance) to notify you and your coworkers of chemical hazards in the workplace. The label also indicates what types of personal protective equipment are needed to provide adequate protection from hazardous materials encountered on the job.

Employers are not required to use the HMIS® labeling system. Any labeling system is fine to use, as long as workers are trained to understand what hazards the label indicates.

HMIS® labels always appear as a rectangle-shaped block of colored bars with hazard ratings of zero through four. Early versions of the HMIS® labels used a blue "health" bar on top, a red "flammability" below that, followed by a yellow "reactivity" bar and a white "PPE" area. There may be additional space on the label for other information, such as the product name, supplemental warnings, manufacturer information, or additional HMIS® information.

There is a new generation of HMIS® being introduced, HMIS® III. This new system still uses colored bars and numeric hazard ratings, but instead of "reactivity" it uses "physical hazard" warnings.

Written Hazard Communication Programs

Your company is required to develop and implement a written hazard communication program. This program details how your company will meet the standard's requirements for labels, MSDSs, and employee information and training.

Your company's written program needs to include:

- A list of the hazardous chemicals present at your site;

- How the MSDS requirements are being met;

- What type of labeling system, if any, is used;

- Detailed information on training compliance;

- Methods used to inform you of the hazards of non-routine tasks and such things as unlabeled piping; and

- Methods used to inform employers of other workers at your site, such as service representatives and subcontractors.

Training

You must be trained at the time of your initial employment or assignment, as well as whenever a new chemical hazard is introduced to your worksite.

According to the hazard communication standard, you are to be informed of the requirements of the standard and any operations in the work area where hazardous chemicals are present. You also need to be informed of the location and availability of your company's written hazard communication program and, more importantly, the location and availability of the MSDS file.

In addition, you will be trained in:

- Methods or observations used to detect the presence or release of hazardous chemicals at your worksite;

- Physical and health hazards of chemicals at your worksite;

- Measures you can take to protect yourself from the hazards, including work practices and personal protective equipment; and

- Details of your employer's hazard communication program, including complete information on labels and MSDSs.

Work at Working Safely

Training is the key to your success and safety as an employee dealing with hazardous chemicals at the worksite. Take it seriously. Get as much as you can from it. Learn about MSDSs, labeling, your company's written program, measures to protect yourself, and what hazardous chemicals you work with. Your good health may depend on how much you learn from your company's training program.

NOTES

Employee _____

Instructor_____

Date _____

Company _____

HAZARD COMMUNICATION REVIEW

1. In the workplace, exposure to chemicals is probably going to be:
   a. Less than at home.
   b. Greater than at home.
   c. For a shorter time period than at home.
   d. To less of a concentration that at home.

2. Containers of hazardous chemicals must be labeled in _____.
   a. Spanish.
   b. English.
   c. The primary language of the employees on the jobsite.
   d. All the above.

3. Training must be provided at least:
   a. At the time of initial employment.
   b. Every six months after that.
   c. Every year.
   d. Once a month.

4. Your company must have a material safety data sheet for each chemical that poses a physical or health hazard at your worksite.
   a. True
   b. False

5. Reactivity, flammability, and health are _____ of chemicals.
   a. Hazard categories.
   b. Chemical properties.
   c. Physical properties.
   d. None of the above.

6. Corrosives are chemicals that:
   a. Cause warts.
   b. Cause cancer.
   c. Eat away at skin.
   d. None of the above.

7. The HMIS® program uses a _____ to notify you and your coworkers of chemical hazards in the workplace.
   a. Numerical hazard rating system.
   b. Symbol hazard rating system.
   c. Graphic hazard rating system.
   d. Pictorial hazard rating system.

8. The HMIS label indicates what types of _____ is/are needed to provide adequate protection from hazardous materials encountered on the job.
   a. Material safety data sheets.
   b. Personal protective equipment.
   c. National Fire Protection labels.
   d. Handling and storage requirements.

9. MSDSs must have all of the following pieces of information except:
   a. Whether the chemical is a carcinogen.
   b. The date of preparation of latest revision.
   c. Identity of chemical as used on its label.
   d. Date the chemical was manufactured.

10. A chronic health hazard:
   a. Occurs immediately.
   b. Occurs over time.
   c. Always causes death.
   d. Never causes death.

HEALTH AND WELLNESS: KEEPING IN PHYSICAL & MENTAL SHAPE

The saying "attitude is everything" does ring true. A positive attitude—plus a healthy lifestyle—goes a long way in keeping you focused on the important things, such as going home safe at the end of each workday. That's why it is so necessary to periodically review your lifestyle to see if any improvements can be made.

Diet

The first place you can start to work on improving your health is your diet. Try to eat a balanced diet. Having a variety of foods will help you get the necessary nutrients and the right amount of calories to maintain a healthy weight.

Many people are trying to lose weight. To do this many of them attempt to modify their diet. However, some diets require you to eat only certain types of foods. Before going on one of these restrictive diets talk to your doctor to see if it is right for you.

Vitamins

Even if you eat right, you still may need to take vitamins or minerals, especially if you are dieting or not eating enough fruits and vegetables. In addition, vitamins and minerals may fight against worksite hazards. For example, vitamins B-6 and E help strengthen tendons and muscles, and help you resist illnesses like carpal tunnel. The mineral, calcium, may strengthen your bones to help avoid fractures if you fall.

Check with your doctor if you think you may need vitamins and minerals. Taking too many can cause toxicity problems and poison your body. So if you do take vitamins and minerals, take only the recommended daily amount (RDA), unless your doctor says otherwise.

Water

In addition to the food and vitamins, water is a key to health and wellness. You need water to digest food, cleanse your body of toxins, and "refuel" yourself after hard work or exercise, especially in extreme heat.

Normally, you should drink at least eight 8-ounce glasses of water per day. However, when you work in extreme heat, you should drink five to seven ounces of water every 15 to 20 minutes. Keep in mind, though, that you can get some of this water through drinks other than water and through fruits and vegetables.

Exercise and Physical Fitness

Exercise is another major ingredient necessary for health. According to the American Heart Association, "up to 250,000 deaths a year in the U.S. can be attributed to a lack of regular physical activity."

While many construction jobs provide plenty of exercise, some jobs like operating a crane or other heavy equipment keep workers in the same position all day. These workers should try to accumulate at least 30 minutes of physical activity each day. Walking, biking, even snow shoveling and washing the car are considered exercise.

If you already get enough physical activity on the job, you may want to consider exercise for a particular part of your body such as your back or your hands. Exercise may strengthen these areas to prevent injuries like carpal tunnel, back pain, and other injuries.

As with any exercise, start slowly and build up to a higher frequency and duration of exercise. Consult your doctor if you have any concerns or known health problems.

Rest

The challenge of just making it through a work shift while fighting fatigue is a battle workers deal with more and more. Fatigue is the condition of being physically or mentally tired or exhausted. Extreme fatigue can lead to uncontrolled and involuntary shutdown of the brain. If you work with machinery, flammable liquids, explosives, hazardous waste, chemicals, electricity, or heights over four feet, or you operate a forklift or other vehicle, the errors caused by fatigue can be critical.

The causes of fatigue include:

Sleep loss	Noise	Drinking too much alcohol
Changes to the body circadian rhythms	Illness	Poor nutrition
Night work	Excessive exposure to toxic chemicals	Not enough exercise
Exertion	Excessive caffeine intake	Boredom
Heat		

Here's how you can fight fatigue:

- Get eight hours of sleep before starting work.

- Sleep at the same time each day. Rotate shifts clockwise (from day to evening to night).

- Take all scheduled work breaks.

- Get acclimatized to working in the heat to avoid heat fatigue. Drink plenty of fluids.

- Look at the health effects listed on the labels or the material safety data sheets for the chemicals you use frequently. Is fatigue a side effect of exposure? Guard against overexposure by using proper protective equipment.

- When trying to sleep during the daytime, find a cool, dark, quiet location. Use earplugs, soft music, or a fan to block out noise.

- See your doctor about sleep disorders and medications for illness.

- Eat a well-balanced diet. Do not eat a lot before bedtime.

- Avoid an excessive intake of caffeine, alcohol, and cigarettes.

- Exercise to maintain stamina.

Mental Health

In addition to physical fitness, you should also consider your mental health and state of mind. Poor mental health can make it difficult to function in your day-to-day work.

Stress

Stress is a physical or mental response to the pressures of an event or factors of living in general. Though we tend to speak of it in a negative context, stress can be positive or negative.

Biologically, when stress occurs, your body releases hormones which speed up your breathing and heart rate, increase your blood sugar levels and blood pressure, and improve blood clotting. Your body gets into a survival mode, readying itself for a physical emergency. This can be a good thing. You have energy and mental agility to get the job done. You are alert and perform well.

As stress continues, your body temporarily adjusts to the stress. If stress is removed during this adjustment period, your body returns to normal. However, if stress goes on for prolonged periods of time, your body fails to adjust and wears out, weakening your defenses to disease. A body cannot run on high speed forever. This can lead to "burnout." Some stress adds challenge, opportunity, and variety to your life. Too much stress can work against you.

Medically, too much stress can cause you to suffer high blood pressure, pain, breathing trouble, cancer, digestive disorders, insomnia, and fatigue. Psychologically, you may suffer frustration, irritability, anger, impatience, worry, a lack of self-confidence, poor listening, and alcohol or drug abuse. To compound matters, your job can be affected too. Stress can lead to accidents, a loss of priorities, rushing, competition, and anger or inappropriate behavior.

Stress Prevention

You can deal with stress in the workplace by watching for the warning signs. Become aware of when you are under stress. Look for signs of being in survival mode. Once you are aware of what stresses you, you can manage your stress by using one or more of the following stress-reduction techniques:

- Take breaks and learn to relax fully.

- Release stress with exercise.

- Maintain proper rest and diet so that you can deal with stressful situations.

- Practice deep breathing to relax.

- Manage your time. Set priorities and do the most important things first.

- Build your self-confidence.

- Share your work if you can't do it all.

- Use laughter as a stress reliever.

- Avoid taking medication or drinking alcohol to eliminate stress temporarily. Your problem will not be solved.

- Talk to a friend about what is bothering you.

Harassment

Another workplace stressor is harassment. Harassment can be racial, sexual, age, nationality, religious, and other types. Unfortunately, victims of harassment often suffer sleeplessness, mood swings, stress, and anxiety. If harassment is a concern at work, know how to handle a harassing co-worker:

- Tell the harasser that you do not welcome his/her behavior and that he/she must stop;

- Report the behavior to a supervisor or manager;

- Follow company procedures to deal with the harassment; and

- Report any repeat or new incidents and whether it came from the same harasser or a new one.

If confronted by an angry or hostile co-worker, stay calm, listen, and watch for a safe chance to escape.

Drug and Alcohol Dependency

Did you know that construction workers have one of the highest rates of illicit drug and alcohol use compared to other industries? Worker impairment caused by mood-altering substances is not new. For several decades alcohol has headed the list of drugs that can adversely impact an employee's health; however, use of illegal drugs like cocaine, amphetamines, and marijuana; over-the-counter medications; as well as abuse of prescription drugs, poses an increasingly large problem at the jobsite.

Drug and alcohol dependency can change your personality, reactions, and judgment. This can cause friction and violence among employees, damage to equipment, poor decisions, absenteeism, and worst of all—accidents. That's why many companies have a drug and alcohol policy, train their supervisors to watch for signs of drug and alcohol use, and start drug testing programs.

Drug Testing Programs

If your company chooses to establish a drug testing program, testing types may include:

- **Pre-employment**—Screening of job applicants;

- **Safety-sensitive**—Testing workers whose jobs involve the safety of co-workers or the public (vehicle operators or flaggers);

- **Reasonable suspicion**—Test persons who appear to be high on drugs;

- **Post-accident**—Test persons involved in an accident that might have been caused by drug use;

- **Return-to-duty**—Testing in connection with return-to-service physical exams;

- **Follow-up**—Testing as a follow-up after treatment;

- **Random**—Random testing to identify drug users and to discourage future use; and

- **Universal**—Testing all workers.

If You Know Someone with a Drug or Alcohol Problem . . .

Tell the person that based on what you've seen, you think something is happening and it concerns you. Urge that person to get help. Some signs of substance abuse are:

- **Physical signs**—Unusual clumsiness and frequent illness.

- **Mood**—Unusually lighthearted one day and very depressed the next.

- **Absenteeism**—More than normal.

- **Actions**—Reacts violently when things go wrong.

- **Accidents**—More accidents.

- **Relationships**—Easily irritated by others. Prefer being by themselves rather than interact with others.

If You Think You Have a Drug or Alcohol Problem . . .

Because this type of problem does not usually go away by itself, you should talk to a friend or loved one. Also, see someone in your company's confidential employee assistance program or call an information hotline: Alcoholics' Anonymous (212-686-1100); the Center for Substance Abuse Treatment (800-662-HELP); 1-800-COCAINE; or the National Clearinghouse for Alcohol and Drug Information (800-729-6686).

Healthy Attitudes

Your health and wellness are affected by not only the worksite but by your attitude. Practicing the following ways of thinking, may keep you healthy and happy.

Think Before You Act

Because carelessness and inattention on the job can lead to disaster, do all you can to think things over before acting. What are you about to do, and what are the consequences? Review the steps for doing your job safely, and use the right tools for the job.

Many work situations don't have listed steps to follow. The only thing you can do then is keep a safety head—one that prepares you for new tasks, new hazards, and new situations. Accidents often occur when you face a new job or a change at work. Think about your safety when something at your jobsite changes. If you are not sure about how to proceed, ask questions. Also, if you discover a hazard, report it.

Even if you are an experienced worker who knows your job well, you can still have an accident if you take safety procedures for granted. From time to time, look over your safety procedures. This way they will be fresh in your mind.

Avoid Shortcuts

Imagine that your table saw has jammed with wood shavings or your blade lost some teeth. Would you unplug the machine from the electrical outlet and apply a "Do Not Plug In" tag before trying to free the shavings or replace the blade? What if you were in a rush to finish the job and you were getting behind already?

Don't let shortcuts tempt you. Imagine that you did not tag the machine and nothing happened. Will you take the shortcut next time? With shortcuts, no longer is it "if" an accident will happen, but "when."

Be a Team Player

Teams tend to work better than one person because team members can pool their skills together. You and your co-workers may know and share safer ways of doing things. You may watch out for each other and save yourselves from injury or illness. And when your team is successful, you feel good knowing you were a part of that success.

Being a team player at a jobsite means you act like part of a team. In addition to getting along, being a team player means:

- Taking responsibility to carry your own weight,

- Motivating and listening to others,

- Cooperating and remaining flexible,

- Getting work done on time,

- Acting professionally, and

- Building respect and trying to understand others.

Work at Working Safely

Good health *can* contribute to reduced injury and illness. So don't take your health, safety, and well-being for granted. Make it a priority.

Employee _____

Instructor_____

Date _____

Company _____

HEALTH AND WELLNESS REVIEW

1. A _____ attitude is an important part of keeping you safe at work.
   a. Negative.
   b. Positive.
   c. Opinionated.
   d. Biased.

2. Before going on a diet that restricts your intake of certain types of foods you should:
   a. Eat a lot of the foods you can't have on the new diet.
   b. Fast for one week.
   c. Talk to your doctor.
   d. None of the above.

3. How many 8-ounce glasses of water should you drink each day?
   a. Three.
   b. Five.
   c. Eight.
   d. Twelve.

4. The absolute minimum amount of exercise a person should get per day is:
   a. 15 minutes
   b. 30 minutes.
   c. 2 hours.
   d. 3 hours.

5. Fatigue is the condition of being physically or _____ tired or exhausted.
   a. Mentally.
   b. Insanely.
   c. Incredibly.
   d. Emotionally.

6. To avoid fatigue on the job:
   a. Use caffeine to stay awake.
   b. Eat lots of carbohydrates.
   c. Get eight hours of sleep per day.
   d. Both a. and b.

7. Stress is a _____ or mental response to the pressures of an event or factors of living in general.
   a. emotional.
   b. physical.
   c. unnatural.
   d. none of the above.

8. If stress goes on for prolonged periods of time, your body defenses become weakened.
   a. True
   b. False

9. If harassment is a concern at work:
   a. Tell the harasser that you do not welcome his/her behavior and that he/she must stop.
   b. Report the behavior to a supervisor or manager.
   c. Follow company procedures to deal with the harassment.
   d. All the above.

10. Which of the following substances can cause workers to be unsafe when working?
    a. Illegal drugs like cocaine, amphetamines, and marijuana.
    b. Over-the-counter medications.
    c. Prescription drugs.
    d. Alcohol.
    e. All the above.

JOBSITE EXPOSURES: AWARENESS IS THE KEY

Protecting yourself against the harmful "stuff" you can breathe or unintentionally eat (like silica dust) or absorb through your skin can be difficult if you don't know what's harmful. That's what this chapter will attempt to do—give you an awareness of some of the most common harmful jobsite exposures and ways to combat them.

Where Are the Regulations?

OSHA regulations for jobsite exposures are found throughout 29 CFR 1926. Here are some common ones:

29 CFR 1926	Name
.53	Ionizing radiation
.54	Nonionizing radiation
.55	Gases, vapors, fumes, dusts, and mists
.60	Methylenedianiline
.62	Lead
.1101-.1152	Toxic and Hazardous Substance (including asbestos, carcinogens, and cadmium)

Asbestos

Asbestos is found in building materials such as shingles, floor tiles, cement pipe, roofing felts, insulation, ceiling tiles, fire-resistant drywall, and acoustical products. Today, fortunately, very few asbestos-containing products are being produced and installed. Consequently, you will find most exposure occurs during the removal of asbestos and the renovation of structures containing asbestos.

If you are not properly protected, your chances of exposing yourself to asbestos are high:

Exposure	Effects
Acute (short term)	Shortness of breath, chest or abdominal pain, or irritation of the skin and mucous membranes
Chronic (long term)	Breathing difficulty, dry cough, broadening and thickening of the ends of the fingers, bluish discoloration of the skin, asbestosis (an emphysema-like condition), lung cancer, and/or mesothelioma (a cancerous tumor that spreads in the lungs and body organs).

Asbestos fibers enter the body by the inhalation or ingestion of airborne particles that become embedded in the tissues of the respiratory or digestive systems. Should exposure occur, asbestos symptoms may not surface for 20 or more years.

Unfortunately, it does not take much asbestos to go over OSHA's exposure limit. Imagine that you had a pinch of asbestos between your thumb and forefinger and you threw it into the air. That's enough to meet the exposure limit.

Protective Measures

There are several ways to control asbestos exposure:

Monitoring: Your company must determine whether an airborne concentration of asbestos is present and whether it exceeds exposure limits specified by OSHA.

Engineering Controls: If exposure is beyond limits specified by OSHA, your company will attempt to minimize the exposure with engineering controls like local exhaust ventilation and enclosing processes which generate asbestos.

Good Work Practices:

- Use the correct, clean, NIOSH-approved respirator. Put on and remove respirators outside the regulated asbestos area. Fit testing guarantees a tight seal.

- Use protective clothing (i.e., full-body suits, gloves, and foot-wear). Store street clothes separately from work clothes. Never wear contaminated clothes home.

- Place and store contaminated clothing in closed containers that prevent the dispersion of the asbestos outside the container.

- Shower and change into clean clothes, including shoes, before leaving the worksite so that no asbestos contamination is carried home.

- Wash your hands and face before eating, drinking, smoking, or applying cosmetics.

- Eat, drink, or smoke in areas outside the work area. Keep all lunch boxes and coffee cups away from the work area. Use a separate lunch area.

Signs: Watch for asbestos danger signs.

Medical Surveillance: Annual medical surveillance of individual workers helps detect employee overexposure and asbestos protection failures.

Asphalt Fumes

Asphalt is a petroleum product used extensively in road paving, roofing, siding, and concrete work. When hot asphalt is applied in a molten state, it generates toxic fumes. Workers exposed to asphalt fumes have reported headaches, breathing problems, and skin irritation. Inconclusive studies have reported lung, stomach, and skin cancers.

There is much debate about whether or not workers are truly exposed to asphalt fume levels high enough to cause concern. OSHA does not yet have an exposure limit for asphalt fumes.

However, if you have side effects from working around asphalt fumes, you may want to wear proper personal protective equipment like respirators, heavy duty gloves, splash goggles or effective safety glasses, long pants and sleeves, and boots. If you are burned by asphalt, apply cold water and get medical attention. If you have breathing trouble, move to an area with fresh air.

Cadmium

OSHA estimates that approximately 70,000 employees in construction are potentially exposed to cadmium. Unfortunately, cadmium exposure can threaten you if you perform construction activities like these without protection:

- Wrecking, demolishing, and salvaging structures where cadmium or cadmium-containing materials are present;

- Cutting, brazing, grinding, or welding on surfaces painted with cadmium-containing paints; and

- Transporting, storing, and disposing of cadmium or cadmium-containing materials on the site.

Let's look at the effects of cadmium exposure:

Exposure	Effects
Short-term high exposure	If ingested—stomach irritation, leading to vomiting and diarrhea. If inhaled—constriction of the throat, chest pain, weakness, fever, severe lung damage, and death.
Long-term low exposure	Build up of cadmium in the kidneys causing kidney disease; lung damage; fragile bones.

When absorbed into the body in certain doses, cadmium is a toxic substance. Cadmium is absorbed by:

Absorption method	Description
Inhalation (breathing)	When cadmium is scattered in the air as dust or fume, it can be inhaled and enter the lungs.
Ingestion (eating)	Cadmium can be absorbed through the digestive system. If a worker handles food, cigarettes, chewing tobacco, or cosmetics that have cadmium on them, or handle them with cadmium-covered hands, cadmium may be ingested.

Eye exposure may cause redness and pain. Skin exposure may result in irritation. In both cases wash with large amounts of water. In all cases of exposure, seek medical attention.

Protective Measures

There are several methods of controlling exposure to cadmium:

Monitoring: Your company will determine whether cadmium is present and whether it exceeds exposure limits specified by OSHA. You will be notified of the results.

Engineering Controls: If exposure is beyond limits specified by OSHA, your company will attempt to minimize the exposure with engineering controls like modifying or enclosing a process which generates cadmium dust or fumes.

Good Work Practices:

- Use the correct, clean, NIOSH-approved respirator. Put on and remove respirators outside the regulated cadmium area. Fit testing guarantees a tight seal.

- Use protective clothing. Store street clothes separately from work clothes. Never wear contaminated clothes home.

- Shower and change into clean clothes, including shoes, before leaving the worksite so that no cadmium contamination is carried home.

- Keep the worksite clean. Use only a vacuum with a HEPA-filter or, if not available, wet sweep, when removing cadmium dust. Never use compressed air for cleaning.

- Wash your hands and face before eating, drinking, smoking, or applying cosmetics.

- Eat, drink, or smoke in areas outside the work area. Keep all lunch boxes and coffee cups away from the work area. Use a separate lunch area.

Signs: Where cadmium hazards are present, you'll see a cadmium danger sign.

Labels: Containers containing cadmium, cadmium compounds, or cadmium-contaminated items must bear a cadmium danger label.

Medical Surveillance: Periodic medical surveillance of individual workers helps detect cadmium protection failures. If cadmium levels are too high, a worker must be temporarily removed from the job to a place of significantly lower exposure without loss of earnings, seniority, rights, or benefits.

Carcinogens

Carcinogens are hazardous substances at your jobsite that can cause cancer. To find out if any of the chemicals you work with are carcinogens, check either the container label or the material safety data sheet for that hazardous substance.

If you work with a carcinogen, your employer may try to reduce your exposure using control measures like ventilation. If control measures do not reduce your exposure enough, you may be required to wear personal protective equipment.

Diesel Exhaust

Thousands of construction workers are exposed daily to diesel exhaust from diesel engines in transportation and off-road vehicles. This exposure increases the risk of adverse health effects including headaches, nausea, cancer, and respiratory disease.

Fortunately, many construction activities are performed outdoors. The great outdoors is a pretty good ventilation system. However, by the same token, many construction activities are also performed indoors, where diesel-fueled equipment is often used. For these areas:

- Use good ventilation.

- Try to run these engines only when necessary in order to cut down on accumulation of diesel exhaust.

- If your equipment uses diesel fuel, check the material safety data sheet (MSDS) for exposure limits, symptoms of exposure, and precautions listed on the MSDS regarding exhaust.

- Report any symptoms of exposure to your supervisor.

Lead

Lead exposure can threaten you if you perform activities like abrasive blasting, sanding, scraping, cutting, burning, welding, and painting during repair, reconstruction, dismantling, and demolition work. If you're not properly protected, your chances of lead poisoning are high:

Exposure	Effects
Short-term large dose	Brain disorder escalating to seizures, coma, and death from cardiorespiratory arrest.
Long-term overexposure	Decreased reaction time, weakness in fingers/wrists/ankles, memory loss, nervous system damage, kidney disease, reproductive impairment, anemia, even death.

When absorbed into the body in certain doses, lead is a toxic substance. Lead is absorbed by:

Absorption method	Description
Inhalation (breathing)	When lead is scattered in the air as a dust, fume, or mist, it can be inhaled and absorbed through the lungs and upper respiratory tract. Operations that generate lead dust and fumes include: flame-torch cutting; welding; the use of heat guns; sanding, scraping and grinding of lead-painted surfaces in repair, reconstruction, dismantling, and demolition work; maintaining process equipment; and exhaust duct work.
Ingestion (eating)	Lead can be absorbed through the digestive system. If a worker handles food, cigarettes, chewing tobacco, or cosmetics that have lead on them, or handles them with lead-covered hands, lead may be ingested.
Skin	Most lead is not absorbed through the skin; however, certain organic lead compounds, such as tetraethyl lead, are ab sorbed this way.

Protective Measures

There are several ways to control lead exposure:

Monitoring: If lead is present at your worksite in any quantity, your company must determine whether exposures reach or exceed OSHA-defined levels.

Engineering Controls: If exposure is beyond the OSHA-defined exposure limit, your company must attempt to minimize the exposure with engineering controls like HEPA vacuums, distancing employees from abrasive blasting operations, chemical removal instead of hand scraping, replacement of lead-based painted building components, brushing/rolling paint instead of spraying, substituting other coatings for lead-based coatings, mobile hydraulic shears cutting instead of torch cutting, and encapsulating lead surfaces.

Good Work Practices:

- Use the correct, clean, NIOSH-approved respirator. Put on and remove respirators outside the lead area. Fit testing guarantees a tight seal.

- Keep the worksite clean. Use only a vacuum with a HEPA-filter or wet-cleaning methods when removing lead dust. Never use compressed air for cleaning.

- Eat, drink, or smoke in areas outside the worksite. Keep all lunch boxes and coffee cups away from the work area. Use a separate lunchroom.

- Wash your hands and face before eating, drinking, smoking, or applying cosmetics.

- Use protective clothing. Store street clothes separately from work clothes. Never wear contaminated clothes home.

- Shower and change into clean clothes, including shoes, before leaving the worksite so that no lead contamination is carried home.

Signs: If exposure exceeds the OSHA-defined exposure limit, you'll see lead warning signs.

Medical Surveillance: Periodic medical surveillance of individual workers helps detect lead protection failures. If a worker's blood lead level is 50 mg/deciliter, then the company must temporarily remove that worker from his/her regular job to a place of significantly lower exposure without loss of earnings, seniority, rights, or benefits.

Mold

Recently, there has been a lot of news about mold damage on construction and renovation sites. However, many people don't understand how or why this construction mold problem is occurring.

Molds can be found almost anywhere; they can grow on virtually any substance when moisture is present. Molds produce tiny spores to reproduce, just as plants produce seeds. These mold spores waft through the indoor and outdoor air continually.

When mold spores land on a damp spot indoors, they may begin growing and digesting whatever they are growing on in order to survive. There are molds that can grow on wood, paper, carpet, foods, even dynamite. When excessive moisture or water accumulates indoors, mold growth often occurs, particularly if the moisture problem remains undiscovered or unaddressed.

There is no practical way to eliminate all mold and mold spores in the environment; the way to control mold growth is to control moisture.

Exposure to Mold

It is estimated that about 50 to 100 common indoor mold types have the potential for creating health problems. Exposure to mold has been identified as a potential cause of many health problems including asthma, sinusitis, and infections. It is also believed that molds play a major role in cases of sick building syndrome and related illnesses.

Mold Damage

The construction jobsite is an environment that is typically ripe for mold growth. Building materials that have been soaked with water, such as fiberglass insulation, wall board, ceiling tiles, and carpeting are excellent media for microbial growth.

Water leakage on furnishings or within building components can result in the proliferation of microorganisms that can release acutely irritating substances into the air. Usually, where microorganisms are allowed to grow, a moldy smell develops. This moldy smell is often associated with microbial contamination and is a result of chemicals released during microbial growth.

Good Work Practices

Three conditions must exist in buildings before microbial contamination can occur:

- High humidity (over 70 percent),

- Appropriate temperatures, and

- Appropriate growth media.

Your company must take steps to make sure these three conditions don't occur. It's important that all potential points of entry for water are inspected for and that any building materials that are used are not wet or moisture-laden.

Radiation

The term radiation connotes danger and death, and it should. While the effect of exposure depends on the type of radiation, the energy, the dose, the quantity, and the part of the body that is exposed, health effects from exposure include nausea, vomiting, diarrhea, weakness, shock, and death.

Furthermore, long-term exposure contributes to an increase in the risk of cancer. However, if you are protected and trained, radiation technology can safely help you do your job. There are two types of radiation:

Radiation type	Includes
Ionizing	Includes alpha rays, beta rays, gamma rays, X-rays, neutrons, high-speed electrons, high-speed protons, and atomic particles.
Nonionizing	Sound or radio waves, visible light, or infrared or ultraviolet light.

At most construction sites the only radiation that is used is non-ionizing radiation produced by laser equipment. Lasers have been helping the construction industry for many years, especially in the area of leveling and alignment applications. New technologies are being developed all the time with many coming in the area of controlling the blades on bulldozers and road graders for precision cutting and grading. However, lasers on the jobsite must be operated only by workers trained in the use of the laser equipment.

Protective Measures

Follow these safety precautions for laser equipment:

- Look for standard laser warning placards (these will be posted in areas in which lasers are used);

- Set up operating laser units above the heads of employees when possible;

- Use beam shutters or caps, or turn the laser off, when laser transmission is not actually required;

- Turn off the laser, when the laser is left unattended for a long time, such as during lunch hour, overnight, or at the change of shifts;

- Use only mechanical or electronic means as a detector for guiding the internal alignment of the laser;

- Look for a laser equipment label which indicates maximum output;

- Never direct the laser beam at workers;

- When it is raining, snowing, dusty, or foggy, do not operate laser systems, as practicable. In any event, keep workers out of range of the area of source and target during such weather conditions; and

- Use anti-laser eye protection if you are working in an area where a potential exposure to direct or reflect laser light greater than five milliwatts exists.

Reproductive Hazards

Because many jobsite chemicals can damage male and female reproductive systems, these chemicals are becoming an increasing concern. Reproductive effects are:

Reduced fertility	Reduced libido	Birth defects
Unsuccessful fertilization or implantation	Menstrual dysfunction	Developmental or behavioral disabilities
	Prenatal death	
An abnormal fetus	Low birth weight	Cancer

Some chemicals that cause these effects are found in use at work, including heavy metals such as lead and cadmium, organic solvents (glycol ethers), and chemical intermediates (styrene and vinyl chloride). While OSHA has no general rules for reproductive hazards, it does regulate four chemicals known to cause reproductive hazards found at many jobsites:

Dibromochloropropane	Cadmium
Lead	Ethylene oxide

Reproductive hazards can enter your body, depending on the substance, by inhalation, skin contact, or ingestion. Therefore, properly wash your hands before eating, drinking, or smoking. Check the material safety data sheet (MSDS) for each chemical you work with to see if it is a reproductive hazard and to find out what other precautions you can take.

Silicosis

Crystalline silica (SiO_2) is a natural compound found in materi-

DANGER

INHALATION HAZARD
DO NOT BREATHE
DUST

als such as concrete, masonry, and rock. Most crystalline silica comes in the form of quartz. Common sand can be as much as 100 percent quartz. When these materials are made into a fine dust and suspended in the air such as during sandblasting, the fine particles produced can cause lung damage, lung cancer, heart failure, and tuberculosis. The lung disease associated with silica dust is called silicosis. About 300 deaths a year are attributed to silicosis.

Early stages of the disease may actually go unnoticed. However, continued exposure may result in:

Shortness of breath progressing to breathing difficulty and respiratory failure

Pain in the chest

Loss of appetite

Occasionally bluish skin at the ear lobes or lips

Fever

Fatigue

Death

Silicosis can be "acute," meaning it can develop after short periods of exposure, or it can be "chronic," meaning it can occur after ten or more years of exposure to lower levels of the dust. See a doctor if you experience any symptoms and suspect that you are exposed to crystalline silica.

In construction, there are a number of activities in which silica dust presents itself:

Sandblasting to remove paint and rust	Dry sweeping or pressurized air blowing of concrete or sand dust
Jackhammer operations	Tunneling operations
Chipping, hammering, drilling, crushing, loading, hauling, and dumping rock	Repairing or replacing of linings of rotary kilns and cupola furnaces
Mixing, chipping, hammering, drilling, sawing, grinding, and demolition of concrete and masonry	Setting, laying, and repairing railroad track

Protective Measures

Keeping dust from getting into the air is the number one way of preventing silicosis. This can be accomplished by simply using a water hose to wet dust down where and when it is created. Other protection methods include:

- Using dust collection systems on dust generating equipment;

- Using a saw that provides water to the blade when sawing concrete or masonry;

- Using water through the drill stem to reduce the amount of dust in air when rock drilling;

- Using local exhaust ventilation to prevent dust from being released into the air;

- Using good work practices to minimize exposures to nearby workers;

- Using abrasives containing less than 1 percent crystalline silica during abrasive blasting to prevent harmful quartz dust from being released in the air;

- Using a respirator when required, not as a primary protection, but when all possible controls are in place and the dust levels are above the National Institute for Occupational Safety and Health (NIOSH) Recommended Exposure Level;

- Measuring dust levels in the air;

- Changing into disposable or washable work clothes at the worksite;

- Showering and changing into clean clothing before leaving the worksite;

- Not eating, drinking, smoking, or applying cosmetics in dust containing areas;

- Washing your hands and face before eating, drinking, smoking, or applying cosmetics; and

- Having a medical examination that includes a chest x-ray, a pulmonary function test, and an annual evaluation for tuberculosis.

Solvents

Many materials you work with everyday contain solvents. Examples of solvents include:

Alcohol	Methylene chloride	Toluene
Benzene	Methyl ethyl ketone	Trichloroethylene
Gasoline	Mineral spirits	Turpentine
Kerosene	Paint	Varnish
Lacquer	Petroleum distillates	Xylene

The health hazards associated with exposure to solvents include:

Nervous system toxicity	Liver and kidney damage	Cancer
		Dermatitis
Reproductive damage	Respiratory impairment	

The material safety data sheet (MSDS) for each solvent you work with will contain information on the health hazards of the solvent. You will find these MSDSs at your jobsite.

Most solvents will enter your body when you breathe them in. That's why it is so important to have adequate ventilation and/or wear a proper respirator. However, a solvent can also enter your body by passing through your skin, and eating food or drinking liquids contaminated with the solvent.

OSHA has established "permissible exposure limits" or PELs for over 100 solvents, including those most commonly used. PELs are amounts of solvent in the air that cannot be exceeded. If your company finds that it exceeds the PEL for a solvent, it must use control measures like ventilation to reduce your exposure.

When these controls are impossible or inadequate, you may be required to wear personal protective equipment, like respirators. Whatever the PEL, always follow safe handling and use procedures for the solvents you work with.

Temperature Extremes

Excessive temperatures can be unbearable for many workers. Heat stress, hypothermia, and frostbite can develop. Heat and cold can even kill. Because of the dangers, it is important for you to understand disorders associated with temperature extremes and what precautions you can use to prevent them at your worksite.

Extreme Heat

Although the body tries to maintain its temperature, hot jobsite conditions can lead to five harmful heat disorders: heat exhaustion, heat stroke, heat cramps, heat syncope (fainting) and heat rash (prickly heat).

Environmental factors affecting the severity of heat disorders include temperature, humidity, radiant heat (heat from the sun or other equipment), and air velocity. The length of exposure and amount of physical activity can affect your condition. Your ability to function in hot conditions is also dependent on age, weight, fitness, diet, medical condition, and acclimatization.

Heat is more likely to affect workers who are not acclimatized to its effects. It takes approximately five to seven days of gradually increased exposure to acclimate to heat. The body changes to make continued exposure to the heat more bearable.

Protective Measures

Some things you can do to prevent heat disorders are:

- Wear loose fitting, light clothing, such as cotton;

- Drink a lot of cool water—5-7 ounces every 15-20 minutes;

- Avoid eating large meals before working in the heat;

- Alternate work tasks to limit time in extreme heat;

- Avoid caffeine and alcoholic beverages—these cause the body to lose water;

- Take frequent, short breaks in a cool, shaded area;

- Use power tools instead of hand tools;

- Wear personal protective equipment (PPE) appropriate for use in the heat (PPE itself should not cause heat stress); and

- Understand heat disorders.

Extreme Cold

When exposed to cold, your body tries to conserve heat for your brain and major organs by reducing circulation to your skin, then to your extremities, and then to organs other than the heart and brain. Eventually you can't move, breathing becomes weakened, your brain loses reasoning power, you become semi-conscious, your heart beats irregularly then stops, and death occurs. Excessive cold can lead to accidents due to forgetfulness, inability to use hands, and exhaustion associated with hypothermia and frostbite.

Environmental factors affecting the severity of cold disorders include temperature, wetness, and wind chill. Additional factors that affect your situation include length of exposure, type of clothing, amount of exposed skin, and body movement. Dehydration, caffeine, alcohol, and some medications can affect your body's ability to detect a change in body temperature.

Protective Measures

Some things you can do to prevent cold disorders are:

- Be aware of the environmental conditions that could lead to cold-related distress;

- Recognize the signs and symptoms of cold-related illnesses;

- Wear the proper clothing; layer clothing to adjust to changing temperatures and weather conditions;

- Wear a hat, gloves, and underwear that keeps moisture away from the body;

- Avoid fatigue or exhaustion; energy is needed to create muscle heat;

- Use the buddy system;

- Drink warm, sweet beverages; avoid caffeine and alcohol;

- Eat warm, high calorie foods;

- Get plenty of rest to help your body cope with the cold;

- Stay dry and remove clothing that gets wet; and

- Check your body for signs of frostbite—especially hands, feet, ears, and face.

Ventilation

Construction work can put all kinds of contaminants in the air. It's the nature of activities like sandblasting, grinding, polishing, buffing, spray finishing, and working over open surface tanks to clean or finish material. Let's look at the various types of contaminants on the next page.

Dusts
Dusts are solid particles generated by handling, crushing, grinding, rapidly impacting, detonating, or heating materials such as rock, cement, metal, coal, or wood. Most construction dusts consist of particles that vary greatly in size, with small particles outnumbering the large ones. When dust is noticeable in the air, there are probably more invisible dust particles present than visible ones. These dusts, especially the smallest ones that are considered respirable, can cause difficulty in breathing and over long-term exposures, respiratory illnesses and death. Silica is a good example of a dust that causes the respiratory illness called silicosis.

Fibers
Fibers are solid particles with a length several times greater than their diameter. Asbestos is one example of this type of fiber. These are major contributors to breathing difficulty and respiratory diseases due to long-term exposure.

Fumes
Fumes are formed when material from a solid condenses in cool air. The solid particles that make up a fume are extremely fine and breathable and potentially harmful. Painting, welding, and other operations involving vapors from molten metals may produce fumes that can be harmful under certain conditions. Because fumes from some of these operations may be toxic, headaches are common symptoms of fume contamination.

Mists

Mists are suspended liquid droplets generated by breaking up a liquid into a dispersed state by splashing or spraying. Sources include oil mists from cutting and grinding operations, acid mists from electroplating, and spray mist from spray finishing operations.

Vapors

Vapors are substances normally solid or liquid at room temperature and pressure. Evaporation changes a liquid to a vapor. Solvents vaporize easily. Solvents with low boiling points can form vapors easily at room temperatures.

Gases

Gases are formless fluids that expand to fill the space to which they are confined. Gases include arc-welding gases, engine exhaust gas (such as from dump trucks and forklifts), and air.

Your company must not allow these contaminants to exceed OSHA-specified limits. It will use various forms of ventilation like exhaust fans, jets, ducts, hoods, separators, etc. to control contaminants. However, where limits are still exceeded, even after ventilation techniques, appropriate respirators are a must.

Protective Measures

How can you protect yourself from these contaminants?

- Know the locations of hazards, including open surface tanks, welding areas, and sandblasting areas. Recognize warning signs of a contaminant problem.

- Use a NIOSH-approved respirator to reduce your exposure and/or provide adequate oxygen. Store it for ready access. Know how to put it on and fit and remove it.

- Wear other proper personal protective equipment. Use rubber boots; glove types and lengths designed for the job; goggles and face shields for chemical splashing; aprons, coats, jackets, sleeves, or other garments made of rubber or materials impervious to liquids.

- Have a standby person outside a tank in the nearest uncontaminated area. He/she must have a suitable respirator and be able to communicate with the employee in the tank. Use a lifeline for confined space and tank work so that the standby person can rescue the employee inside.

Work at Working Safely

While there are a number of jobsite exposures to be aware of, it may be easier to think of these exposures as any harmful things you:

- Breathe,

- Unintentionally eat,

- Absorb through your skin, and

- Endure (heat and cold).

It makes sense then that most good work practices will prevent you from breathing, eating, absorbing, or enduring the hazard. Understanding this connection between exposures and work practices will help keep you safe.

NOTES

Employee _____

Instructor _____

Date _____

Company _____

JOBSITE EXPOSURES REVIEW

1. Three diseases associated with asbestos exposure include:
   a. Melanoma, asbestosis, and lung cancer
   b. Asbestosis, lung cancer, and mesothelioma
   c. Asbestosis, lung cancer, and tremolite
   d. Tremolite, lung cancer, and asbestosis

2. Area(s) of the body most directly affected by cadmium include:
   a. Heart.
   b. Lungs and kidneys.
   c. None. Cadmium is safe.
   d. Central nervous system including the brain.

3. Lead exposure can threaten you if you perform activities like:
   a. Abrasive blasting.
   b. Sanding and scraping.
   c. Welding.
   d. All the above.

4. When mold spores land on a _____ spot indoors, they may begin growing and digesting whatever they are growing on in order to survive.
   a. Damp.
   b. Warm.
   c. Slippery.
   d. None of the above.

5. Silicosis, the disease associated with exposure to silica dust, can be:
   a. Acute.
   b. Chronic.
   c. Both a. and b.
   d. None of the above.

6. Substances that cause reproductive hazards will have that information listed on the MSDS.
   a. True.
   b. False.

7. It takes _____ days of gradually increased exposure to acclimate a person to heat.
   a. One to two.
   b. Three to four.
   c. Five to seven.
   d. Ten.

8. When exposed to extreme cold, your body tries to conserve heat by reducing circulation to your:
   a. Brain.
   b. Skin.
   c. Heart.
   d. None of the above.

9. Some things you can do to prevent heat-related disorders are:
   a. Avoid caffeine and alcoholic beverages.
   b. Drink very limited amounts of water.
   c. Wear tight clothing, preferably black in color.
   d. All the above.

10. Some ways to prevent cold-related disorders are:
    a. Leave wet clothing on.
    b. Do not eat much.
    c. Layer your clothing.
    d. Work alone.

LADDERS & STAIRWAYS: CLIMBING WITH SAFETY

Ladders and stairways are a major source of accidents at construction sites. In 2002 there were 126 fatalities and 24,086 injuries resulting from falls from ladders. There were also 11 fatalities and 20,507 injuries from falling down stairs or steps.

Where Are the Regulations?

The regulations for ladders and stairways are found at 29 CFR 1926.1050-.1060.

General Requirements for Stairways & Ladders

- Your employer must provide a stairway or ladder at all access points with a difference in elevation of 19 inches or more and a ramp, runway, sloped embankment, or personnel hoist is not provided.

- Stairway and ladder safety requirements must be in place before you begin work.

- Access points from one level to the next must be kept clear to permit free passage of employees.

- Two or more access points must be provided when one is restricted because of work or equipment.

Ladders

A ladder can be a great help on your job. Ladders generally can be placed into three categories: fixed, portable, and job-made. When you are choosing what portable ladder to use, consider the:

- **Type**—Step ladders are good for work close to the ground, whereas, straight or extension ladders are good for higher work.

- **Length**—Choose the right length so work can be done from a convenient height without lots of reaching or working from an unsafe position.

- **Material**—Wood is nonconductive, heavy, hard to move, and hard to inspect for rot. Metal is light and weather resistant, but conducts electricity. Fiberglass is light, long-lasting, nonconductive, and has good traction.

- **Strength**—Choose light, medium, or heavy-duty strength depending on the weight and stress the ladder has to bear.

While ladders are uncomplicated and simple to use, you shouldn't take ladder safety for granted.

All ladders must:

- Have surfaces that prevent splinters, snagging, or other injuries.

- Be free of oil, grease, or other slipping hazards.

- Have nonconductive siderails if used where you or the ladder could contact energized electrical parts.

- Must be inspected by a competent person for visible defects periodically and after an event that could affect the safe use of the ladder. If defects like corrosion or broken, faulty, or missing parts are found, the ladder must not be used until repairs are made. Tag a defective ladder "DO NOT USE" or clearly identify it as defective.

- Have rungs, cleats, and steps that are parallel, level, and uniformly spaced when the ladder is in use.

- Be placed on stable and level surfaces unless "tied off" to prevent accidental movement. Do not place a ladder on slippery surfaces unless it has slip-resistant feet or is secured.

However, slip-resistant feet are not a substitute for proper placement, lashing, or holding the ladder on a slippery surface. Both are better.

- Never be tied or fastened to another ladder to make it longer.
- Have clearance around the top and bottom of the ladder.
- Not be moved, shifted, or extended while someone is on it.

When going up or down a ladder you must face the ladder and use at least one hand to grasp the ladder at all times. Do not carry anything that could cause you to lose your balance and fall.

Portable Ladders

When a portable ladder is used, place the angle of the ladder so that the ladder's base is one foot out for each four feet of ladder working height (support point to base).

The ladder siderails must extend at least three feet above the upper landing surface. If this is not possible because of the ladder's length, the ladder must be "tied off" at the top to a rigid support that will not bend. Also, a grabrail at the top must be provided to help employees on and off the ladder.

Fixed Ladders

Fixed ladders cannot be readily moved or carried because they are an integral part of a building or structure. These ladders must:

- Have cages, wells, ladder safety devices, or self-retracting lifelines when the climb is less than 24 feet but the top of the ladder is at a distance greater than 24 feet above lower levels.
- Have the following where the total length of a climb equals or is greater than 24 feet:
  - Ladder safety devices; or
  - Self-retracting lifelines, and rest platforms at intervals not to exceed 150 feet; or

- A cage or well, and multiple ladder sections, each ladder section not to exceed 50 feet in length. Ladder sections must be offset from adjacent sections, and landing platforms must be provided at maximum intervals of 50 feet.

Ladder safety devices, and related support systems for fixed ladders must:

- Permit you to go up and down without continually having to hold, push, or pull any part of the device, leaving both hands free for climbing.

- Be activated within two feet after a fall occurs.

Job-Made Ladders

Job-made ladders must be constructed for their intended use. Cleats must be spaced between 10 and 14 inches apart and be uniform.

Wood job-made ladders with spliced siderails must be used at an angle such that the horizontal distance is one-eighth the working length of the ladder.

Stairways

We all have misstepped when climbing or descending stairs. In fact, one study claimed that people misstep on stairs about every 2,000 steps. Unfortunately, missteps can turn into falls. Poor traction and tripping over objects placed on stairs also leads to falls.

Because we use stairs so often, it's easy to forget they can be hazardous. Here's what you can do to protect yourself from injury:

- Use handrails whenever possible. If you are carrying something, use extra caution.

- Don't run up or down stairs or jump from landing to landing.

- Don't carry a load that you can't see over.

- Report any unsafe conditions promptly. Maybe you can't control lighting or a cluttered stairway, but you can report them to your supervisor or maintenance staff.

- Report broken stair treads, floor boards, or handrails.

General requirements for stairways include:

- Stairways, not a permanent part of a structure being worked on, must have landings of not less than 30 inches in the direction of travel and extend at least 22 inches in width at every 12 feet or less of vertical rise.

- Where doors or gates open directly on a stairway, a platform must be provided. The swing of the door must leave 20 inches of effective width.

- All parts of stairways must be free of hazardous projections such as protruding nails.

- Slippery conditions must be eliminated before use.

Stairrails and Handrails

- Stairways having four or more risers, or rising more than 30 inches, whichever is less, must be equipped with at least one handrail and one stairrail system along each unprotected side or edge.

- Stairrails must not be less than 36 inches high measured from the tread to the top of the stairrail.

- Handrails must be between 30 and 37 inches high measured from the tread surface.

- Handrails and the top rails of stairrail systems must be able to withstand a 200 pound force applied within two inches of the top edge, in any downward or outward direction, at any point along the top edge.

- Stairrail and handrail surfaces must prevent splinters, lacerations, or snagging of clothing.

- Handrails must provide an adequate handhold to avoid falling. If the handrails are not permanent, there must be at least a three inch clearance between the handrail and wall.

- Unprotected sides and edges of stairway landings must be provided with guardrail systems. Guardrails must meet the requirements of the fall protection rule.

Ladder and Stairs Training

You must be trained in the safe use of ladders and stairways. The training must be given by a competent person and train you to:

- Recognize hazards related to ladders and stairways and ways to minimize the hazards;

- Understand the proper construction, use, placement and care in handling stairways and ladders;

- Know the maximum load ability for stairways and ladders; and

- Understand the OSHA regulations for ladders.

You will be retrained, as often as necessary, to make sure your understanding of stairway and ladder safety remains sharp.

Work at Working Safely

It is fairly easy to find construction sites that do not take ladder and stairway safety measures. In fact, the ones most often missed are listed below:

- Ladder siderails must extend at least three feet above the upper landing surface. If this is not possible because of the ladder's length, the ladder must be "tied off" at the top to a rigid support that will not bend. A grabrail at the top must be provided to help employees on and off the ladder.

- Stairways having four or more risers or rising more than 30 inches, whichever is less, must be equipped with at least one handrail, and one stairrail system along each unprotected side or edge.

- Your employer must provide a stairway or ladder at all access points with a difference in elevation of 19 inches or more and no ramp, runway, sloped embankment or personnel hoist is provided.

- Ladders with structural defects, such as broken or missing rungs, cleats, or steps, broken or split rails, corroded components, or other faulty or defective components must be immediately tagged and withdrawn from use until repaired.

Look around at your site. Maybe you can point out problems and prevent a serious accident.

NOTES

Employee _____

Instructor_____

Date _____

Company _____

LADDERS & STAIRWAYS REVIEW

1. Ladders or stairways must be provided when the difference in elevation is _____ inches.
    a. 12.
    b. 15.
    c. 19.
    d. 21.

2. All ladders must be inspected by a _____ person for visible defects periodically and after an event that could affect the safe use of the ladder.
    a. Qualified.
    b. Competent.
    c. Trained.
    d. None of the above.

3. You must tag a defective ladder:
    a. "Out of service".
    b. "Not safe".
    c. "Do not use".
    d. "Return to manufacturer".

4. When a portable ladder is used, the angle must be _____ foot out for each four feet of ladder working height.
    a. One.
    b. Two.
    c. Three.
    d. None of the above.

5. Ladder safety devices for fixed ladders must be activated within _____ feet after a fall occurs.
    a. One.
    b. Two.
    c. Three.
    d. None of the above.

6. Landings are required every 12 feet or less for stairways that are not a permanent part of a structure.
   a. True.
   b. False.

7. Where doors open directly on a stairway platform, the swing of the door must leave _____ inches of effective width.
   a. 10.
   b. 20.
   c. 30.
   d. 40.

8. Job-made ladders must be constructed for their intended use and have cleats spaced between _____.
   a. 8 and 10 inches.
   b. 10 and 12 inches.
   c. 10 and 14 inches.
   d. 12 and 15 inches.

9. Handrails and stair top rails must be able to withstand a _____ pound force applied in various directions.
   a. 100.
   b. 150.
   c. 200.
   d. 250.

10. Ladder and stairway training must be given:
    a. Before first use.
    b. As often as necessary.
    c. After an accident.
    d. Both a and b.

LOCKOUT/TAGOUT: CONTROL HAZARDOUS ENERGY

The accidental release of hazardous energy can cause accidents and injuries. Lockout/tagout procedures prevent you from accidentally being exposed to injurious and even life-threatening situations from energized equipment or circuits.

Where Are the Regulations?

OSHA regulates lockout/tagout procedures at 29 CFR 1926.417—Lockout and Tagging of Circuits. This is part of the electrical section of the construction standards. Lockout/tagout procedures can also be found in 29 CFR 1926.702—Requirements for Equipment and Tools (Concrete and Masonry Construction).

What Is Lockout/Tagout?

Lockout is the process of turning off and locking out the flow of energy from a power source to a piece of equipment or circuit, and keeping it locked out.

Lockout is accomplished by installing a lockout device at the power source so that equipment powered by that source cannot be operated. A lockout device is a lock, block, or chain that keeps a switch, valve, or lever in the off position.

Locks are provided by your employer and can be used only for lockout purposes. Never use these locks for tool boxes, storage sheds, or other devices.

Tagout is placing a tag on the power source. The tag acts as a warning not to restore energy—it is not a physical restraint. Tags must clearly state, "Do not operate or remove this tag," or the like, and must be applied by hand.

Both locks and tags must be strong enough to prevent unauthorized removal and to withstand various environmental conditions.

What Must Be Locked or Tagged Out?

A lock and a tag must be placed on each handle, circuit breaker, push button, or similar device used to deenergize electric equipment when repairs or maintenance is done.

In general, all power sources that can be locked out must be locked out for servicing or maintenance. However, guards or interlock devices cannot be used as substitutes for locks during major servicing.

OSHA rules require your employer to:

• Maintain a written copy of the lockout/tagout procedures for your company and make it available to you;

• Instruct you to recognize and avoid unsafe conditions that require lockout and/or tagout; and

• Instruct you in the OSHA lockout/tagout rules that apply to your work.

Controlling Energy Sources

A wide variety of energy sources require lockout/tagout to protect you from the release of hazardous energy. Some of these energy sources include: electrical, mechanical, pneumatic, hydraulic, chemical, and thermal sources.

Some of the problems an accidental release of hazardous energy could cause are: (1) accidental start-ups, (2) electric shock, and (3) release of stored, residual, or potential energy. These accidents often occur when someone takes a short cut during machinery servicing, or when workers don't understand the equipment or the lockout/tagout procedures for the job.

Lockout/Tagout Procedures

Although lockout/tagout is specifically mentioned in the construction regulations at just a few places, those procedures can be applied to all electrical and mechanical lockout/tagout situations at your worksite. It would be foolish to just apply them to electrical and concrete work.

Electrical Equipment Lockout/Tagout

Anytime electrical equipment is deactivated for repair, or circuits are shut off, the equipment must be tagged at the point where it can be energized. This is done to warn anyone that there is maintenance work going on.

The procedure outlined here must be followed in the order given to accomplish lockout and tagout.

General Requirements

• Disconnect the circuits and equipment requiring work from all electrical energy sources.

• Do not use control circuit devices, such as push buttons, selector switches, and interlocks as the only means for deenergizing circuits or equipment.

• Do not use interlocks for electric equipment as a substitute for lockout and tagging procedures.

- Deenergize stored electric energy, such as capacitors.

Application of Locks and Tags

- Place a lock and tag on each disconnecting means (switches, breakers, push buttons, levers) used to deenergize equipment and circuits on which work is to be performed.

  - If a lock cannot be applied, or your employer can demonstrate that tagging will provide a level of safety equivalent to that of a lock, a tag may be used without a lock.

  - A tag used without a lock must have at least one other safety measure that provides a level of safety equal to a lock. Examples include: the removal of an isolating circuit, blocking of a controlling switch, or opening of an extra disconnecting device.

  - A lock may be placed without a tag when all of the following conditions exist: (1) only one circuit or piece of equipment is deenergized; and (2) the lockout period does not extend beyond the workshift; and (3) employees exposed to the hazards associated with the reenergizing of the circuit or equipment are familiar with the procedure.

- Attach the lock so as to prevent others from operating the switch unless they resort to undue force or the use of tools.

- Ensure each tag has a statement prohibiting unauthorized operation of the switch and removal of the tag.

- Before any circuit or equipment can be considered deenergized and work can proceed, qualified person must:

- Operate the equipment controls or otherwise verify the equipment is deenergized, and

- Use test equipment to ensure electrical circuits are deenergized.

Qualified persons are those who, by education, or professional standing, or who by extensive knowledge, training, and experience, have successfully demonstrated their ability to solve or resolve problems relating to the subject matter, the work, or the project.

Before you reenergize deenergized equipment, a qualified person must conduct tests and visual inspections as necessary to verify that all tools, and electrical jumpers, are removed.

Each lock and tag must be removed by the employee who applied it or under their supervision. If this employee is absent, their lock or tag may be removed by a qualified person provided that:

- The employer ensures that the employee who applied the lock or tag is not available at the workplace; and

- There shall be a visual determination that all employees are clear of the circuits and equipment.

Mechanical Equipment Lockout/Tagout

You cannot perform maintenance or repair on mechanical equipment such as compressors, mixers, or pumps used for concrete and masonry construction activities, where the accidental operation of the equipment could occur and cause injury.

After servicing of electrical or mechanical equipment is finished, replace all machine guards and make sure all tools are removed from the area. Only then can you remove your tag and lock and reconnect all sources of energy. After this, you may restart the equipment according to the procedures outlined here and in the OSHA construction regulations.

Other Concerns

Other concerns that must be addressed in your company's lockout/tagout program include:

- **Outside Contractors**—If your company acts as an outside contractor, you must be informed of the host company's lockout or tagout procedures. That way you'll understand the meaning of locks or tags as you come across them while working. The host company must also be aware of your procedures so their employees are aware of the meanings under your program.

- **Personnel Changes**—In general, if a piece of equipment is locked out at a personnel change, the person beginning the job must apply his/her lock before the employee who is leaving the job can remove his/hers. This procedure ensures that lockout or tagout protection is kept.

- **Power Sources That Cannot Be Locked Out**—In very rare cases, a power source cannot be physically locked out. Discuss this situation with your supervisor to find out if tagout alone may be used safely. There are a few situations where tagout alone is allowed.

Work at Working Safely

Always follow lockout/tagout procedures during servicing or maintenance of equipment, where unexpected energization or start-up of the equipment could harm you or a fellow employee. When you service or repair electric equipment, always lock and tag out power sources and switches. Never ignore or remove the locks or tags of other employees, and know your role as an authorized employee. Your attention to, and respect for, your company's lockout/tagout program will make your worksite safer for both you and your coworkers.

Employee _____

Instructor_____

Date _____

Company _____

LOCKOUT/TAGOUT REVIEW

1. A lockout device is a lock, block, or chain that keeps a switch, valve, or lever in the _____ position.
   a. Intermediate.
   b. On.
   c. Off.
   d. Inverse.

2. Tagout used alone is _____ locking the equipment out.
   a. As effective as.
   b. Less effective than.
   c. More effective than.
   d. Safer than just.

3. Anytime electrical equipment is deactivated for repair, or circuits are shut off, the equipment must be tagged at the point where it can be:
   a. Maintained.
   b. Locked.
   c. Lubricated.
   d. Energized.

4. After servicing of electrical or mechanical equipment is finished:
   a. Remove all tools from the area.
   b. Remove your lock and tag.
   c. Reconnect all sources of energy.
   d. All the above.

5. _____ use control circuit devices, such as push buttons, selector switches, and interlocks as the only means for deenergizing circuits or equipment.
   a. Always.
   b. Never.
   c. It's OK to sometimes.

6. A tag used as a lock must have _____ other safety measure that provides a level of safety equal to a lock.
   a. One.
   b. Two.
   c. Three.
   d. Four.

7. If the employee that applied the lock or tag is absent, his/her lock can be removed, in certain circumstances, by a(n):
   a. Authorized person.
   b. Qualified person.
   c. Supervisor.
   d. Safety manager.

8. There are a few situations where tagout alone is allowed.
   a. True
   a. False

9. After servicing of electrical or mechanical equipment is finished, replace all:
   a. Tools.
   b. Tags.
   c. Machine guards.
   d. Locks.

10. If your company acts as an outside contractor:
   a. You must be informed of the host company's lockout/tagout procedures.
   b. You can prevent the host's employees from locking out equipment.
   c. You must provide the host company with your lockout/tagout procedures.
   d. Both a. and c.

MATERIALS HANDLING & STORAGE: MOVING EARTH AND MATERIALS

For construction, materials handling equipment covers a wide range of jobsite workhorses. OSHA divides materials handling equipment into two groups:

> **Earthmoving equipment**—Primarily moves dirt around, but some can double as materials haulers. Examples include scrapers, loaders, crawlers, wheel tractors, bulldozers, off-highway trucks, graders, agricultural and industrial tractors, and similar equipment.

> **Lifting and hauling equipment**—Primarily moves raw materials around your jobsite. Examples include lift trucks, stackers, forklifts and other powered industrial trucks, handlers, helicopters, and similar equipment.

We have also decided to cover:

> **Rigging equipment**—Used with cranes, derricks, hoists, and other properly rigged equipment to move materials at jobsites.

While this equipment makes life easier at the construction site, it can also be a source of frustration and injury when not used properly. Materials handling and storage equipment kills hundreds and injures thousands of workers each year. One of the worst kinds of accidents occurs when employees fall from moving equipment and are run over, ending up in the hospital or morgue.

Where Are the Regulations?

OSHA regulations for materials handling and storage are scattered throughout 29 CFR 1926:

29 CFR 1926	Name
.20	General Safety and Health Provisions
.250 -.252	Materials Handling, Storage, Use, and Disposal
.550 -.556	Cranes, Derricks, Hoists, Elevators, and Conveyors
.600 -.606	Motor Vehicles, Mechanized Equipment, and Marine Operations
.1000 -.1003	Rollover Protective Structures; Overhead Protection

Inspections

Your equipment is only as safe as your mechanical inspections. It is required that you inspect your equipment before you start your shift; that way if you find something wrong, you can report it to your supervisor and avoid potential accidents.

Pre-Operational Inspection

Many things can make equipment unsafe. That is why a pre-operational (pre-shift) walk around is important. You should always perform a pre-operational inspection on equipment you plan to operate. The inspection could include checking the following:

- Access ladders, steps, guardrails, and hand-holds are securely fastened and in good condition;

- Walking and/or stepping surfaces are free of debris and/or slippery substances and that non-skid surfaces, where required, are in good shape;

- Brakes and clutches for adjustment and operation;

- Outriggers when used;

- Boom hoist lockout on cranes;

- Other operator aids such as crane anti-two-block devices and load moment indicators;

- All electrical distribution and transmission lines are de-energized and visibly grounded, insulating barriers are erected to prevent physical contact with the lines, or OSHA-required clearances are followed;

- Barricades are set up to protect employees from being struck or crushed by the equipment;

- Tires, hoses, belts, and cables are in proper order;

- Lift trucks, stackers, etc., have the rated capacity clearly posted;

- Forklift mast and forks are in good shape;

- No leaks, missing bolts, or broken welds; and

- Slings, fastenings, and attachments do not have damage or defects.

Cab Inspection

You should check the cab before starting your equipment to ensure:

- Door latches work and doors can be opened from the inside or outside;

- Seat suspensions operate properly;

- Seat belts are provided;

- Forward and rear adjustment, weight compensators, and other adjustable features work properly;

- Employee transport vehicles have a firmly attached seat for each passenger;

- A proper, full-pressure fire extinguisher is accessible;

- Your field of vision is not obstructed;

- Appropriate placards are posted if required;

- A load rating chart is available if required; and

- All bi-directional machines, such as rollers, compactors, front-end loaders, bulldozers, are equipped with an audible horn, distinguishable above the construction noise.

After start-up, check all gauges and warning lights for proper readings. Operate all controls to ensure they are functioning properly.

The inspection lists above are not exhaustive. Be sure to follow the checklist provided by your company and/or equipment manufacturer.

Rigging Inspection

Slings can be deadly if overloaded, allowed to deteriorate, or not inspected regularly. A competent person must inspect a sling before it is used and as often as necessary during use.

OSHA defines competent person as "one who is capable of identifying existing and predictable hazards in the surroundings or working conditions which are unsanitary, hazardous, or dangerous to employees, and who has authorization to take prompt corrective measures to eliminate them."

At a minimum, learn to recognize dangerous slings. If you find one, don't use it.

Mounting & Dismounting Equipment

Many workdays are lost because of slips, trips, and falls on and from heavy equipment. However there are proper ways of climbing up, getting into or out of cabs and seats, and getting down and off of equipment.

You should:

- Always use handrails, face the ladder or steps, and maintain three points of contact (two feet and one hand, or one foot and two hands) at all times;

- Always use anti-slip surfaces for climbing or stepping; and

- Never jump from or to ladders, steps, or walkways.

Operating Equipment

Whether you are an equipment operator, spotter, or someone working in the vicinity of materials handling and storage operations, you must always be aware of what is going on and know where your coworkers are

as you go about your work. Never allow others to hitch a ride on equipment, unless they are authorized and there is a safe seat to ride in. Only operate equipment if you are qualified by training or experience and designated by your company to operate that type of equipment. And never park in another vehicle's blind area.

Earthmoving Equipment

The following list provides some general operating rules for earthmoving equipment:

- Be sure the service brakes can stop and hold the equipment when fully loaded;

- During normal operation, guard all hazardous scissors points on front-end loaders;

- Do not drive any equipment or vehicles on any access roadway or grade unless the roadway/grade is made to handle the equipment or vehicles involved;

- Ensure that every emergency access ramp and berm is constructed to restrain and control runaway vehicles;

- Use seat belts when provided; and

- Operate the horn as needed when the machine is moving in either direction.

Lifting and Hauling Equipment

Here are some general operating rules for lifting and hauling equipment:

- Keep clear of loads being lifted and suspended loads;

- Do not exceed ratings; and

- Completely lower any load engaging means, neutralize controls, and set the brakes to prevent movement when you are about to leave your equipment unattended (operator is 25 or more feet away).

Cranes

The crane, a type of lifting and hauling equipment, has its own specific requirements in addition to the general ones covered previously:

- Do not estimate your load weight. Know what it is and don't exceed the crane's capacity;

- Ensure all calculations regarding a lift are correct;

- When lowering your boom, ensure you do not lower it too far and thereby shift the crane's center of gravity;

- Know the manufacturer's operating instructions and limitations;

- Don't work with a damaged crane;

- Don't operate carelessly. The slightest touch of the boom on an object such as a wall, or a load hitting the boom, can cause the boom to collapse;

- Use your signal person when necessary;

- Prior to a lift ensure the ground is not soft and is reasonably level;

- Ensure correct-sized outrigger floats are used;

- Do not allow anyone except the oiler, instructor, or competent person to be on the equipment when it is in operation. Do not hoist, lower, swing, or travel while anyone is on the load or hook; and

- Be careful during dismantling procedures—booms can be damaged and collapse during dismantling operations.

The OSHA regulations do not require that crane operators be "certified" to operate cranes, but you must be trained to operate them safely. Many states and cities do require certification. OSHA is working on a regulation that may require certification.

Forklifts

Driving a construction forklift is fundamentally different than driving cars or other trucks. Forklifts tend to steer more easily loaded than empty and are driven in reverse as often as forward. Unlike cars, forklifts must have excellent stability for operating in dirt, mud, snow, and up and down grades and slopes. However, most forklifts have a greater chance of tipping over when suddenly turned or when moving loads.

Another difference with fork-lifts at the jobsite is that they may have different steering modes. Your forklift may have all-wheel, front-wheel, or rear-wheel steering. It may also be steered by locking one set of wheels. Know what the manufacturer's requirements are and follow them.

Although fundamentally different than cars, forklifts *like* cars can have collisions with property and people. Wear your seat belt and follow these rules for preventing collision:

- Stay alert;

- Slow down and sound the horn at corners;

- Slow down on slippery surfaces and watch for oil and grease spots;

- If you cannot see past a load in front, travel backwards, carefully, or don't proceed without a signal person;

- Face the direction of travel;

- Be familiar with the manufac-turer's performance specifi-cations for the truck;

- Follow all manufacturer pro-cedures for handling a forklift tipover (this procedure differs per forklift type);

- Know the braking capability (stopping distance, maxi-mum grade, and gears/speed/grade information);

- Know the traffic patterns in use; and

- Give pedestrians/emergency vehicles the right of way.

Forklifts can lift only so much. Check the rating plate for your forklift's load capacity. Each manufacturer provides an identification plate on every forklift they build.

Loads need special care so that they do not fall. In order to prevent tipping and load falling hazards, do the following:

While lifting loads:

- Lift only load weights your forklift can safely handle,

- Chock the wheels of the delivery truck when loading or unloading the truck,

- Use a forking system which suits the load,

- Place the heaviest part of the load closest to the backrest,

- Slide forks or other load engaging means as far as possible under the load,

- Support and center the load so that it will not fall forward or sideways,

- Exercise caution with off-center loads, and

- Tie together unstable loads.

While carrying loads:

- Do not attempt to move loads with broken pallets,

- Tilt the mast back to stabilize the load before moving,

- Travel with the load lowered, and

- Raise and lower loads only when the forklift is stopped.

Most forklifts operate with highly flammable and combustible fuels and battery acid. Many forklifts haul flammable and combustible hazardous materials. Some companies have flammable and combustible chemical storage areas. For these reasons, use the following precautions:

- Never smoke in fueling areas, near flammable and combustible loads, or while charging the battery;

- Prevent open flames, sparks, or electric arcs while fueling, handling flammable and combustible loads, and charging the battery; and

- Never fuel a forklift with the engine running.

Gasoline-powered forklifts give off deadly carbon monoxide exhaust. Only drive this kind of forklift in well-ventilated areas.

Because forklifts are typically used near pedestrians, it is important for both pedestrian and forklift operator to watch out for each other. Although the forklift operator must give pedestrians the right of way, he/she also must not allow riders on any forklift, horseplay around forklifts, nor pedestrians to walk under loads.

Rigging Equipment

Slings are an integral part of materials handling. For your protection, slings must:

- Not be loaded in excess of their safe working loads;

- Balance loads to prevent slippage in a basket hitch;

- Not be shortened with knots, bolts, or other devices;

- Not be used if sling legs have been kinked; and

- Be padded or protected from a load's sharp edges.

Keeping slings in good condition is also important. When they are not in use remove slings from the immediate area so that they are:

BASKET   CHOKER   VERTICAL

- Protected from hazards like dirt, oil, and grease;

- Not a tripping or swinging hazard; and

- Not walked on or run over by construction traffic.

Other rigging safety rules include:

- Don't place your hands/fingers between a sling and its load while the sling is being tightened around the load.

- Keep suspended loads clear of all obstructions.

- Avoid sudden starts and stops when moving loads.

- Remain clear of loads about to be lifted and suspended. Use tag lines when appropriate.

- Never "shock load."

Overhead Obstacles

Falling Objects

Protect yourself from falling objects by:

- Wearing your hard hat;

- Cribbing and blocking equipment that is suspended or held aloft by slings, hoists, or jacks; and

- Fully lowering or blocking equipment when repairing them or finished using them. All controls must be neutral, with the motors stopped and brakes set, unless the work being done requires otherwise.

Overhead Power Lines

Contact between large materials handling equipment and overhead lines is a major cause of fatal occupational injuries in the United States. It is responsible for approximately 5.5 percent of all fatal construction-related injuries each year.

Any overhead wire must be considered energized unless and until the person owning the line or the electric utility authorities indicate that it is not energized and has visibly grounded it. OSHA allows three methods to protect equipment operators from contacting live overhead lines:

- De-energize lines and visibly ground them at the point of work;

- Erect insulating barriers, not a part of or attached to the equipment; and

- Operate equipment or machines according to required clearances between lines and parts of equipment. When operating near power lines, minimum clearance between lines and the equipment or load must be:

Voltage	Minimum Clearance
50 kV or below	10 feet
Over 50 kV	10 feet plus 0.4 inch for each 1 kV over 50 kV, or twice the length of the line insulator but never less than 10 feet.

When in transit with no load and boom lowered, the equipment clearance must be:

Voltage	Minimum Clearance
Less than 50 kV	4 feet
Over 50 kV up to and including 345 kV	10 feet
Over 345 kV up to and including 750 kV	16 feet

A "spotter" must observe line clearance of the equipment and give a timely warning for all operations where it is difficult for the operator to maintain clearance by visual means.

Maintenance and Repair

Do not operate any equipment or rigging with deficiencies or defective parts. If you discover damaged or defective equipment or rigging, immediately remove them from service and tag them as damaged or defective and don't use them until they are repaired or replaced.

No modifications or additions that affect the capacity or safe operation of the equipment can be made without the manufacturer's written approval. The original safety factor of the equipment may not be reduced.

Storage

Movement and storage of materials—providing a continuous flow of raw materials and parts—is vital to any project. Storing materials involves hoisting, carrying, stacking, and other often times dangerous activities. You can be injured by improperly lifting materials (both manually and by machine), falling objects, and improperly stacked supplies.

Try to stack, rack, block, interlock, or otherwise secure all stored materials to prevent them from falling, sliding, or collapsing. If you're storing things higher than the ground level, check to see what the maximum safe load limits (in pounds per square foot) for the given storage area are. Most load limits will be posted in those areas. If not, find out. Then don't exceed these maximum safe load limits.

There are a couple distance requirements for storage:

Don't store materials within	Of
6 feet	Any hoistway or inside floor opening.
10 feet	Any exterior wall that doesn't extend above the top of the material stored.

Specific types of storage also have requirements:

Storage material	Requirement
Bricks	Do not stack over 7 feet. If a loose brick stack reaches 4 feet, taper it back 2 inches for every foot over the 4 feet level.
Masonry blocks	If stacked higher than 6 feet, taper back one-half block per tier over the 6-foot level.
Lumber	Withdraw all nails before stacking. Stack so pile is stable and self-supporting. Do not pile lumber over 20 feet. If you will handle lumber manually, do not stack it over 16 feet.
Structural steel, poles, pipe, bar stock, & other cylindrical materials	Stack and block so as to prevent spreading or tilting, unless you rack them.

Keep storage areas free from accumulation of materials that constitute hazards from tripping, fire, explosion, or pest harborage. That includes keeping clear aisles and controlling vegetation. Do not store materials on scaffolds or runways unless they are for current projects. And lastly, segregate noncompatible materials in storage.

Work at Working Safely

Materials handling and storage equipment are the primary workhorses of the construction industry. Yet they may pose dangers when not operated or maintained properly. Know your equipment, be aware of other equipment around you, and follow your company's storage guidelines.

MATERIALS & HANDLING

Employee _____

Instructor_____

Date _____

Company _____

MATERIALS HANDLING REVIEW

1. Equipment inspections are required _____ your shift.
   a. After.
   b. Before.
   c. During.
   d. Multiple times throughout.

2. Your pre-operational inspection will depend on the type of equipment you will be operating.
   a. True.
   b. False.

3. If a seat belt is provided on earthmoving equipment you do not have to use it.
   a. True.
   b. False.

4. A _____ must inspect a sling before it is used.
   a. Authorized person.
   b. Competent person.
   c. Supervisor.
   d. Safety manager.

5. Lifting and hauling equipment is considered unattended when the operator is ____ or more feet away.
   a. 100.
   b. 75.
   c. 50.
   d. 25.

6. The following people *should not* be on a crane when it is in operation:
   a. Inspector.
   b. Oiler.
   c. Instructor.
   d. Competent person.

7. When operating a forklift:
   a. Never smoke in fueling areas.
   b. Place the heaviest part of the load closest to the backrest.
   c. Know the braking capability.
   d. All the above.

8. Forklift operators must:
   a. Allow riders on the forklift.
   b. Give pedestrians the right of way.
   c. Let pedestrians walk under loads.
   d. Allow horseplay around forklifts.

9. If you discover defective or damaged equipment or rigging, immediately:
   a. Remove them from service.
   b. Tag them as damaged.
   c. Don't use them until repaired.
   d. All the above.

10. If handling lumber manually do not stack it over:
    a. 7 feet
    b. 12 feet.
    c. 16 feet.
    d. 20 feet.

MOTOR VEHICLES: GETTING FROM HERE TO THERE

Construction jobsites are dangerous places. When motor vehicles are operating, they become even more dangerous. However, dangerous does not have to mean unsafe. Proper vehicle inspection and operation can

contribute to making a jobsite a safe place to work.

What Is a Motor Vehicle?

Motor vehicles are one set of heavy equipment operating at jobsites. The other set is materials handling equipment—covered in the Materials Handling and Storage chapter.

> **Motor vehicles**—Those pieces of equipment that can operate both at a jobsite and on the highway. Examples include dump trucks, flatbeds, and pickups. Transporting workers, equipment, and materials are some of their functions.

Where Are the Regulations?

OSHA regulations governing motor vehicles in construction are found under both 29 CFR 1926.600 and 1926.601. These regulations cover required equipment and inspections. There is also a motor vehicle operation requirement found in 29 CFR 1926.20.

Department of Transportation (DOT) regulations are found at 49 CFR 350 to 399. These are the Federal Motor Carrier Safety Regulations which include safety requirements for the driver, his/her vehicle, and its cargo. Hazardous Materials Regulations are found at 49 CFR 171-180.

Required Equipment

Does your vehicle have the following required equipment? If it's missing something, let your supervisor know. Do not operate your vehicle without this equipment:

- Good working service, emergency, and parking brake system.

- At least two operational headlights and two taillights when you need additional light.

- Operable brake lights and a horn or other audible warning device at the operator's station.

- Windshields and powered wipers. Cracked or broken glass must be replaced.

- Seat belts and anchorages.

- Fenders, or mud flaps if not designed for fenders.

There are additional requirements for specific kinds of motor vehicles:

Vehicle	Must have
Dumping	• A latch or other device on all operating levers controlling dumping devices in order to prevent accidental starting or tripping of the mechanism; • Dump truck tailgate trip handles arranged so that in dumping, the operator will be in the clear; and • Support, permanently attached and capable of being locked in position to prevent accidental lowering of the dumping body.
Hoisting	A latch or other device on all operating levers controlling hoisting devices in order to prevent accidental starting or tripping of the mechanism.
Haulage	Cab shield and/or canopy adequate to protect the operator from shifting or falling materials if the pay load is loaded by means of cranes, power shovels, loaders, or similar equipment.
Personnel	A firmly secured and adequate number of seats.

Similarly, no driver may operate a motor vehicle with an obstructed view to the rear, unless either:

- It has a reverse signal alarm that you can hear above the surrounding noise level, or

- The vehicle is backed up when an observer says it is safe to do so.

Inspections

Your motor vehicle is only as safe as your mechanical inspections. Never use any vehicle not in compliance with OSHA and manufacturer standards. Identify a damaged or defective vehicle as unsafe and tag, lock, or physically remove it from the jobsite.

Be aware of whether the vehicle ran normally during the last shift. Check the logbook. Are there deficiencies that need to be corrected?

Always perform your pre-operational inspection with care. Look for damage that could cause an accident or failure while in use. At the **beginning of each shift**, check each of the following items on the motor vehicle you intend to operate:

- All brake systems (service, trailer, parking, emergency)
- Horn
- Coupling devices
- Operating controls
- Tires
- Steering mechanism
- Seat belts
- Safety devices

When such equipment is necessary, these requirements also apply to:

- Lights
- Windshield wipers
- Fire extinguishers
- Reflectors
- Defrosters

You should also add items found on both the manufacturer's and your company's pre-operational checklist. All defects must be corrected before the vehicle is placed back into service. Be familiar with the vehicle's maintenance requirements and records. Find out if the necessary maintenance has been performed.

Also it is a good idea to be aware of your vehicle's blind spots. Check these blind spots during your inspection. That way you'll avoid running into something.

Vehicle Entry and Exit

Believe it or not, there are proper ways of climbing up, getting into and out of cabs and seats, and getting down from motor vehicles:

- Always use handrails, face the ladder or steps, and maintain three points of contact (two feet and one hand, or one foot and two hands) at all times;

- Always use anti-slip surfaces for climbing or stepping; and

- Never jump from or to ladders, steps, or walkways.

Operation

You must be qualified by training or experience to operate motor vehicles. Driving a construction motor vehicle can be very different than driving a car. Here are some general operational rules:

- Sound your horn and follow established procedures before moving the vehicle;

- Know jobsite and highway traffic patterns; and

- Secure all tools and material to prevent movement when transported in the same compartment with employees.

What driving rules do you follow?

When driving your motor vehicle	Follow these rules
On public roads	• State and local laws, and • Dept. of Transportation (DOT) rules.
On the jobsite	• OSHA rules, • Your company rules, and • Good common sense rules.

Gasoline-powered motor vehicles give off deadly carbon monoxide exhaust. Only drive this kind of vehicle in well-ventilated areas.

Hazardous Materials Transportation

As a construction company employee you may be involved in transporting hazardous materials (hazmat). This could be anything from explosives to small or limited quantities of some material declared hazardous by the DOT.

The DOT has identified nine classes of materials that pose unreasonable risk to health and safety/property when transported. These classes are:

Class 1 – Explosives

Class 2 – Gases

Class 3 – Flammables or Combustibles

Class 4 – Flammable Solids

Class 5 – Oxidizers or Organic Peroxides

Class 6 – Poisons

Class 7 – Radioactives

Class 8 – Corrosives

Class 9 – Miscellaneous

Being a hazmat driver requires training in:

- The general provisions of the Hazardous Materials Regulations;

- Hazardous materials recognition and identification;

- Specific requirements applicable to your job functions;

- Emergency response information, self-protection measures, and accident prevention methods and procedures;

- Basic and in-depth security awareness training.

- Handling hazardous spills or leaks; and

- Hazard communication training (see chapter by that name).

Think about your motor vehicles. Do any of them carry aerosol cans of products such as paint, lubricants, pesticides, starting fluid, etc., or small quantities of flammable liquids such as solvents, or "fuses" or "power device cartridges"? How about gasoline for the operation of auxiliary equipment such as power saws and portable generators or oxygen and acetylene for welding? All are hazardous materials subject to the hazmat requirements.

Hazmat requirements include:

- Determining the proper hazard classification,
- Correctly marking packages,
- Preparing shipping papers,
- Placarding vehicles, and
- Being prepared for any situation while transporting materials to the site.

Parking

General parking recommendations are as follows:

- Know the safe parking procedures for your vehicle.
- Don't park in another vehicle's blind area.
- Because mechanical failures can happen, when you are about to leave a vehicle unattended (when you will be 25 or more feet away), it's a good idea to neutralize controls, set the brakes, and shut off the engine to prevent movement.

Work at Working Safely

The bottom line is knowing your vehicle. Be a skilled and safe driver:

- Give pedestrians the right of way,
- Do not allow riders unless the vehicle is designed for passengers,
- Never fuel a motor vehicle with the engine running,
- Never smoke in fueling or battery charging areas, and
- Do not allow horseplay around motor vehicles.

NOTES

Employee _____

Instructor _____

Date _____

Company _____

MOTOR VEHICLES REVIEW

1. All motor vehicles need to have certain minimum equipment requirements that include:
   a. One working headlight (only when additional light is needed).
   b. One working brake light (only when additional light is needed).
   c. Seat belts and seat belt anchorages.
   d. Both a. and b.

2. No driver may operate a motor vehicle with an obstructed view to the rear, unless:
   a. The vehicle has a reverse signal alarm that can be heard above the surrounding noise level.
   b. An observer says it is ok to do so.
   c. Neither of the above.
   d. Both a. and b.

3. Perform your preoperational checklist:
   a. At the end if each shift.
   b. At the beginning of each shift.
   c. At the most convenient time during the shift.
   d. After a breakdown.

4. Which of the following is an item you *do not* check during the preoperational checklist:
   a. Tires.
   b. Safety devices.
   c. Paint.
   d. Horn.

5. When entering or getting out of construction motor vehicle:
   a. Look around before you jump down.
   b. Avoid using the ladder or steps.
   c. Avoid getting out in a puddle.
   d. Use handrails, anti-slip strips, and three points of contact.

6. You must be qualified by _____ to operate motor vehicles.
   a. Your supervisor.
   b. Training or experience.
   c. OSHA.
   d. The local law enforcement agency.

7. On public roads you must follow:
   a. State and local laws.
   b. OSHA rules.
   c. Department of Transportation rules.
   d. Both a. and c.

8. There are _____ classes of materials that pose unreasonable risk to health and human safety/property when transported.
   a. Six.
   b. Seven.
   c. Eight.
   d. Nine.

9. Hazard communication is a required part of the hazmat driver's training.
   a. True.
   b. False.

10. You must set the brakes if you are going to be _____ feet away from a vehicle.
    a. 15 feet.
    b. 25 feet.
    c. 45 feet.
    d. None of the above.

PERSONAL PROTECTIVE EQUIPMENT: YOUR LINE OF DEFENSE AGAINST INJURY

According to the Bureau of Labor Statistics, 226,800 construction workers suffered serious, nonfatal, on-the-job injuries and illnesses in 2002. In addition, over 1,000 construction workers die each year from injuries sustained on the job. That's about three fatalities a day.

These sobering statistics demonstrate that many workers face unsafe conditions or work practices every day. While employers need to minimize these hazards as much as possible at the source, this is not always easy to do. The use of personal protective equipment complements other measures your employer takes to create a safe work environment for you.

Eye Protection

Each year thousands of disabling eye injuries occur in the construction industry. The main cause of job related eye injuries is objects striking a worker's eye. Contact with chemicals also accounts for a large number of injuries.

Eye Protection Regulations

Rules for eye protection are found in 29 CFR 1926.102. This regulation requires employees to use eye protection to guard against injury in situations where injury is likely. The protection equipment must meet the requirements specified in American National Standard Institute (ANSI) Z87.1-1968, *Practice for Occupational and Educational Eye and Face Protection.*

Eye and Face Hazards

Most workers who have had eye injuries were not wearing eye protection. They said that eye protection was not normally used, or they felt it wasn't needed.

Most eye injuries can be prevented by following safety precautions and wearing proper protective equipment. Some of the reported causes of eye injuries are:

- Injurious gases, vapors, and liquids.
- Dusts or powders, fumes and mists.
- Flying objects or particles.
- Splashing metals.
- Thermal and radiation hazards such as heat, glare, ultraviolet, and infrared rays.
- Lasers.
- Electrical hazards.

How Can You Protect Your Eyes?

The first steps to preventing eye injuries are to install equipment guards, have good ventilation and lighting, and use personal protective equipment. Your employer must also provide eyewashes to minimize damage once an injury has occurred.

Eyewash Facilities

Eyewash facilities can include eyewash fountains, drench showers, hand-held drench hoses, and emergency bottles. Very simply, they all use large amounts of water to flush away eye contaminants.

The location of an eyewash facility is very important; your eyes can be damaged very quickly by some contaminants. The first 15 seconds after an injury is the critical period. Therefore, eyewashes should be within 100 feet, or a 10-second walk of the hazardous work area.

Personal Protective Equipment for the Eye

A wide variety of safety equipment is available to keep you safe and injury free. Protective eye and face equipment must comply with ANSI guidelines and be marked directly on the piece of equipment (e.g. glasses frames and lenses).

Safety Glasses

The most common type of protective equipment for the eyes is safety glasses. They may look like normal streetwear glasses, but they have much stronger lenses, are impact resistant, and come in prescription or nonprescription forms. Safety frames are heat-resistant and stronger than streetwear frames. They also help  prevent lenses from being pushed into your eyes.

Safety glasses also are available with side shield guards. Semi-side shields provide protection for the sides of your eyes. Eye-cup side shields provide more thorough eye protection from hazards that come from the front, side, top, or bottom.

Goggles

Goggles are very similar to safety glasses but fit closer to the eyes. They can provide additional protection in hazardous situations involving liquid splashes, fumes, vapors, and dust. Some models can be worn over prescription glasses and others are made with fabric eye cups to provide better ventilation.

Face Shields

Full-face protection is often required to guard against molten metal and chemical splashes. Face shields are available to fit over a hard hat or to wear directly on the head. A face shield should always be used with other eye protection such as goggles or glasses.

What about Contact Lenses?

Most workers can safely wear their contacts on the job. Situations where contacts should be worn with caution include those where you might be exposed to chemical fumes, vapors or splashes, intense heat, and molten metals.

It is important to remember that, if hazards warrant, your contacts should be worn along with additional eye protection. Contacts should be removed immediately if redness of the eye, blurring of vision, or pain, develops on the job.

It's also a good idea to keep a spare pair of contacts or prescription glasses with you in case the pair you usually wear is lost or damaged while you're working. You might also want to make sure your supervisor or first aid providers know that you wear contacts, in the event of any injury on the job.

Care for Your Eye Protection

Your face and eye protection equipment must be kept clean and in good repair. The use of broken or visually defective face and eye protection is prohibited.

Foot Protection

Every day hundreds of workers in the United States suffer disabling injuries to their feet and toes. Foot and toe injuries in the construction industry numbered 8,381 for 2002, according to a Bureau of Labor Statistics report. This number represents over six percent of all disabling injuries. The foot is especially vulnerable to injury. Yet many workers ignore the serious hazards in the workplace and refuse to wear protective footwear.

Foot Protection Regulations

The OSHA regulations for the construction industry are located in 29 CFR 1926.96. OSHA requires that safety toe shoes meet the requirements of the *American National Standards for Men's Safety-Toe Footwear* (ANSI Z41.1-1967).

Foot Hazards

Your feet are vulnerable to many types of skin diseases, cuts, punctures, burns, sprains, and fractures. But sharp or heavy objects falling on the foot are the primary source of injury.

Other hazards include:

- **Compression**—The foot or toe is squeezed between two objects or rolled over.

- **Puncture**—A sharp object like a nail breaks through the sole.

- **Electricity**—A hazard in jobs where workers use power tools or electric equipment.

- **Slipping**—Contact with surface hazards like oil, water, or chemicals causes falls.

- **Chemicals**—Chemicals and solvents corrode ordinary safety shoes and can harm your feet.

- **Extreme heat or cold**—Insulation or ventilation is required depending on climate.

- **Wetness**—The primary hazard may be slipping but discomfort and fungal infections can occur if your feet are wet for long periods of time.

Personal Protective Equipment for the Foot

Foot protection is guarding your toes, ankles, and feet from injury. Protective footwear comes in many varieties to suit very specific work applications:

- **Safety Shoes**—Standard safety shoes have toes that meet testing requirements found in the ANSI standard. Steel, reinforced plastic, and hard rubber are used for safety toes, depending on their intended use. These shoes are worn in many construction jobs.

- **Sole Puncture Resistant Footwear**—Puncture-resistant soles in safety shoes protect against hazards of stepping on sharp objects that can penetrate standard shoe soles. They are used primarily in general construction work.

- **Metatarsal Guards**—Shoes with metatarsal or instep guards protect the upper foot from impacts. In these shoes, metal guards extend over the foot rather than just over the toes.

- **Conductive Shoes**—These shoes permit static electricity that builds up in the body of the wearer to drain off harmlessly to the ground. By preventing accumulation of static electricity, most conductive shoes keep electrostatic discharge from igniting sensitive explosive mixtures. These shoes are often worn by workers in munitions facilities or refineries. Do not use these shoes if you work near open electrical circuits.

- **Safety Boots**—Rubber or plastic safety boots offer protection against oil, water, acids, corrosives, and other industrial chemicals. They are also available with features like steel-toe caps, puncture-resistant insoles, and metatarsal guards. Some rubber boots are made to be pulled over regular safety shoes.

- **Electrical Hazard Shoes**—These shoes offer protection against shock hazards from contact with open circuits of 600 volts or less under dry conditions. They are used in areas where employees work on live or potentially live electrical circuits. The toebox is insulated from the shoe so there is no exposed metal. These shoes are most effective when dry and in good repair.

- **Static Dissipative Shoes**—These shoes are designed to reduce accumulation of excess static electricity. They conduct body charge to ground while maintaining a sufficiently high level of resistance to protect you from electrical shock due to live electrical circuits.

- **Add-On Foot Protection**—Metatarsal guards and shoe covers can be attached to shoes for greater protection from falling objects. Strap-on wooden-soled sandals can be used for protection against the underfoot hazards of oils, acids, hot water, caustics, or sharp objects. Rubber spats protect feet and ankles against chemicals. Puncture-proof inserts made of steel can be slipped into shoes to protect against underfoot hazards. Strap-on cleats fastened to your shoes will provide greater traction.

Hand Protection

How would you answer the question, "What is the most used tool in construction?" Some people would name a commonly used hand tool like a hammer or screwdriver. Others might respond with a list of larger equipment such as lathes or power tools. But the correct answer is deceptively simple. The most used tool in almost any workplace is the human hand.

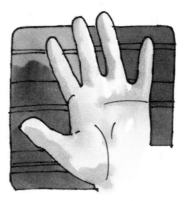

Think of almost any job at your worksite, from sweeping up construction debris to skillfully using a finishing trowel. Your hands and fingers are the tools you use every day. Try writing without using your thumb. Try holding a hammer with only two fingers. Hand protection is important because our hands are exposed to so many hazards in the workplace.

Hand Protection Regulations

There are no safety requirements for hand protection in the construction rules. However, OSHA does regulate hand protection in general industry at 29 CFR 1910.138. We will use those rules as a basis for our discussion here.

Your employer should select and require you to use hand protection when you are exposed to hazards such as skin absorption of harmful substances, severe cuts or lacerations, severe abrasions, punctures, chemical burns, or harmful temperature extremes.

Hand Hazards

At work, your hands are exposed to three kinds of hazards:

- **Mechanical hazards**—These are present wherever machinery is used. Injuries resulting from machinery use might include cuts, punctures, abrasions, or crushing.

- **Environmental hazards**—Factors like extreme heat or cold, electricity, and materials handling have the potential to injure your hands.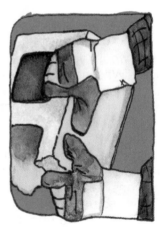

- **Irritating substances**—Skin conditions like dermatitis can be caused by contact with chemicals and biological agents (bacteria, fungi, and viruses). Chemicals can also enter the bloodstream through abrasions or cuts.

The First Defense Against Hand Injuries

The first defense in the battle to reduce hand injuries is engineering controls designed into equipment during manufacture or used to alter the work environment to make it safe and hazard free. Machine guards protect hands and fingers from moving parts and should not be altered or removed. Jobs should be designed to incorporate proper positions for tools, hands, and materials.

Personal Protective Equipment for Hands

Personal protective equipment (PPE) can help reduce the frequency and severity of hand and finger injury. Although fingers are harder to protect, they can be shielded from many common injuries. Personal protection is available in the form of gloves, mitts, finger cots, thimbles, hand pads, sleeves and hand lotions or barrier creams.

Gloves

Gloves are perhaps the most commonly used type of PPE. They provide protection to fingers, hands, and sometimes wrists and forearms. Ideally, gloves should be designed to protect against specific hazards of a job being performed. Types range from common canvas work gloves to highly specialized gloves used in specific tasks.

Good examples of job-rated hand protection are the items designed for those who work with electricity—special rubber gloves and lineworkers' rubber insulating sleeves. The gloves are made of natural or synthetic rubber and are color coded to correspond with their level of voltage protection.

Rubber, vinyl, or neoprene gloves are also used when handling caustic chemicals like acids, cleansers, or petroleum products.

Leather gloves or leather reinforced with metal stitching are useful for handling rough or abrasive materials. Metal mesh gloves are worn by workers who work with sharp knives.

Many gloves are rated as being safe for use with certain kinds of chemicals. If you are allowed to select your own PPE, read the glove manufacturer's chemical resistance charts. They rate each glove material and how it withstands specific chemicals.

Wear only gloves that fit your hand. Gloves that are too small can tire your hands and gloves that are too large are clumsy to work with. Gloves should be worn with great caution near moving equipment or machinery parts. A glove can get caught and pull your fingers or hand into the machinery. Gloves should be given proper care and cleaning. Inspect them regularly for change in shape, hardening, stretching, or rips.

Other PPE for the Hand

There are many other types of hand protection:

- **Mitts** are similar to gloves, but have a division for the thumb and one for fingers.

- **Finger cots** provide protection for a single finger or fingertip.

- **Thimbles** protect the thumb or the thumb and first two fingers.

- **Hand pads** protect the palm from cuts and friction. These pads also protect against burns. Heavier and less flexible than gloves or mitts, they should not be used for jobs requiring manual dexterity.

- **Sleeves or forearm cuffs** protect the wrists and arms against heat, splashing liquids, impacts, or cuts.

Barrier Creams

Barrier creams or lotions can be used by themselves or along with other types of PPE. You might use a lotion when other types of protection cannot be used, such as when working with or near moving machinery. Three types of cream are available:

- **Vanishing cream** usually contains soap/emollients to coat the skin, make cleanup easy, and protect against mild acids.

- **Water-repellent cream** leaves an insoluble film on the skin. It protects against irritants in water—alkalis and acids.

- **Solvent-repellent cream** protects against irritating solvents and oils.

To be effective, creams or lotions should be applied frequently. Remember that these creams do not protect against highly corrosive substances.

What Happens If You Injured Your Hand?

If you do injure your hand, you should know what to do. For cuts, control the bleeding with direct pressure on the wound. For broken bones, immobilize the injured hand. For chemical or heat burns, put the hand under running water and flush for 10 to 20 minutes. Some chemicals react with water so read warning labels on containers or MSDSs for first aid information.

If you are faced with handling a more serious hand or finger injury like an amputation, act quickly. Severed limbs can often be reattached. Control heavy bleeding or shock first. Keep the severed part cool, but don't freeze it. Do not apply a tourniquet unless you have received training in how and when to do so.

No matter what kind of injury occurs, get medical help as soon as possible. Report the accident to your supervisor and take the victim to your company doctor, first-aid station, or hospital emergency room.

Head Protection

Each year there are thousands of head injuries in the construction industry. Injuries range from major concussion to death, minor abrasions to trauma, or even electrocution.

Head Protection Regulations

Head protection regulations for the construction industry are found in 29 CFR 1926.100. The standards recognized by OSHA for protective hats are in the ANSI *Safety Requirements for Industrial Head Protection*, Z89.1-1969. Helmets for employees exposed to high voltage electrical shock and burns must meet the requirements found in ANSI, Z89.2-1971.

Head Hazards

Head injuries are caused by falling or flying objects or by bumping your head against a fixed object. Other head injuries come from electrical shock and burns. Hard hats are designed to do two things: resist penetration, and absorb the shock of a blow. They lessen injury because they are designed with a hard outer shell and a suspension system inside.

When you are working in an area where there is a possible danger of head injury from impact, or from falling or flying objects, or from electrical shock and burns, you must wear your hard hat.

Personal Protective Equipment for the Head

Hard hats fall into two types and three classes and are intended to provide protection against a specific hazardous condition. The types include:

- **Type 1**—Full brim, at least 1 $^1/_4$ inches wide.

- **Type 2**—No brim, peak extending forward from the crown.

The classes of hard hats are:

- **Class A**—For general service, these hats provide impact and limited voltage protection. Mining, building construction, tunneling, and lumbering are examples of industries that use Class A hard hats.

- **Class B**—For utility service, this hat protects your head from impact and penetration from falling and flying objects and high-voltage shock and burn. It is mainly used during electrical work.

- **Class C**—For special service, this hat is designed for light-weight comfort and impact protection. They are used where there is no danger from electrical hazards.

Care for Your Hard Hat

You should take proper care of your hard hat to prolong its life and your safety:

- Check your hat daily for signs of dents, cracks, or penetration. *Do not use* it if any of these signs are found on the shell, suspension, headband, or sweatband.

- Do not put your hat on the rear-window shelf of a car. Sunlight and heat can damage your hard hat.

- Clean your hat once a month in warm, soapy water. Scrub and rinse the shell with clear, hot water.

- Do not paint your hard hat. Some type of paints and thinners may damage the shell or weaken the hard hat itself.

Hearing Protection

When noise at your worksite is above the exposure levels listed in 29 CFR 1926.52, Permissible Noise Exposures, hearing protective devices must be provided and used. Ask your supervisor which site noises exceed permissible exposures.

Hearing protection that is inserted in your ear must fit. Plain cotton is not an acceptable protective device.

Respiratory Protection

When you're working with hazardous dust, smoke, mist, fumes, sprays, vapors, or gases, you can't always see or smell the hazard. That's why your company monitors these hazards and uses control measures like ventilation to minimize your exposure to these substances. However, control measures are not always enough to contain the hazards completely. Fortunately, an appropriate respirator can protect you from hazards in the air you breathe.

Respiratory Protection Regulations

OSHA has issued regulations at 29 CFR 1926.103 governing the use of respirators for construction. Your company is responsible for determining if respirators are needed at the jobsite. If you do need to wear a respirator, your company will also select and provide one for you.

Types of Respirators

There are different makes and styles of respirators, but all of them fall into one of two types: air-purifying and atmosphere-supplying. Once it's determined that respirators are required, your company will develop a work-site-specific respiratory protection program that includes information on respirator selection, medical evaluations, fit testing, use, schedules for respirator care, and procedures to ensure the effectiveness of the program.

Air-Purifying Respirators

These respirators simply remove the contaminants from the air as you breathe. There must be safe levels of oxygen in the air when using an air-purifying respirator. Typically, they have a tight-fitting facepiece and use a filter, cartridge, or canister that is approved for the contaminant that you are exposed to.

Atmosphere-Supplying Respirators

Atmosphere-supplying respirators provide you with breathing air from a clean source. Supplied-air respirators (SARs) use an airline to feed clean air from the source to the respirator's facepiece, helmet, or hood. A self-contained breathing apparatus (SCBA) supplies air from tanks that are carried by the user.

Selecting a Respirator

It is critical that your company match the hazards of the job to the capabilities and limitations of the respirator, because using the wrong respirator could kill you.

If there is not enough oxygen, or the contaminant (or its levels) is unknown, the atmosphere is considered immediately dangerous to life or health (IDLH). Atmosphere-supplying respirators are appropriate for IDLH atmospheres.

However, most atmospheres that require respiratory protection contain enough oxygen and the contaminant levels are known. These are not non-IDLH atmospheres.

Medical Evaluations Are First

Wearing a respirator can be physically taxing to your body. Before you can be required to wear a respirator, you must be evaluated to ensure that you are physically capable to do so. You will need to fill out a medical questionnaire and may need an exam and medical tests. Periodic follow-up medical evaluations may also be required if you exhibit medical signs or symptoms that would have an effect on your ability to wear a respirator. Some examples include shortness of breath, dizziness, chest pain, lung disease, and heart conditions.

How to Fit a Respirator

In addition to the medical evaluation, you must pass a fit test before you can use a respirator with a tight-fitting facepiece. Because one respirator will not fit everyone, you and your company will need to find a respirator size and style that fits you. Without a proper fit, the respirator will not provide protection.

Some people's faces just don't fit well into a respirator facepiece. Generally, if you wear glasses or have a beard, mustache, or long sideburns, your respirator will need some modifications to fit properly.

All fit tests are done using the same type and size of respirator you will be wearing on the jobsite. You will be re-tested at least annually to ensure the fit is still correct.

Using Your Respirator

Each time you put on your respirator you must perform one of two checks to make sure that your respirator has a good face to facepiece seal:

Check	Description
Positive pressure	Close the exhalation valve and breathe out gently into the face piece. The seal is good if you feel the facepiece bulge out slightly without air leaking out around the seal.
Negative pressure	Close the exhalation valves and breathe in gently so the face piece collapses slightly. Hold your breath for ten seconds. The seal is good if the facepiece remains slightly collapsed and there is no air leaking in around the seal.

Whenever you are wearing a respirator, make sure that you always leave the respirator-use area if:

- You detect vapor or gas breakthrough, changes in breathing resistance, or facepiece leakage;

- You need to change filter, cartridge, or canister elements;

- You need to wash your face or the facepiece to prevent eye or skin irritation; or

- Your respirator needs repair.

Care and Maintenance of Respirators

Keep your respirator clean and in good repair. Don't risk irritations, disease, or contamination from using a dirty or damaged respirator. A respirator's cleaning and inspection schedule depends on who uses the respirator and how it is used. Follow the cleaning and inspection frequencies listed in the table below:

If respirator is:	Clean it:	Inspect it:
Your own	As often as necessary	Before each use and during cleaning
Shared with co-workers	Before anyone else wears it	Before each use and during cleaning
Emergency use	After each use	At least monthly

In addition, SCBAs must be inspected monthly. Also follow filter, cartridge, and canister change schedules.

Store your respirator so it is protected from damage, contamination, dust, sunlight, extreme temperatures, excessive moisture, and damaging chemicals.

If your company allows you to wear a respirator on a voluntary basis (when the contaminants are at safe levels), you still need a medical evaluation and must follow your program's provisions for cleaning, storage, and maintaining these respirators. However, no program is needed if only dust masks are used on a volunteer basis. Also, your employer will provide you with basic information on respirator use whether it's voluntary or not.

Work at Working Safely

Your employer can provide you with proper PPE but cannot watch you all day to ensure you use it. PPE can be uncomfortable, cumbersome, and hot, but those are only inconveniences compared to injury. It is your responsibility to work safely so you go home to your family uninjured. Remember these safety tips before beginning each workday:

- Personally owned equipment you use for work must meet the same OSHA requirements as equipment provided by your company.

- Match your safety equipment to the degree of hazards.

- Maintain your equipment in a reliable and sanitary condition and replace it if defective.

- Make sure any safety device you use fits properly.

- Never use street-wear eyeglasses and boots for protection. They were not designed for safety.

- Never store your hard hat in your car window.

- Don't take a chance with your eyes, feet, head, or hearing. "It can't happen to me" is a dangerous thought that has been proven wrong again and again. Obtain proper protective equipment and wear it at all times on the job.

Employee _____

Instructor_____

Date _____

Company _____

PERSONAL PROTECTIVE EQUIPMENT REVIEW

1. Eye protection equipment must meet the safety requirements specified by the:
   a. Department of Transportation.
   b. American National Standards Institute.
   c. Occupational Health and Safety Agency.
   d. Department of Homeland Security.

2. Most workers who have had eye injuries:
   a. Were not wearing eye protection.
   b. Were wearing contact lenses.
   c. Were wearing eye protection.
   d. Never wore eye protection.

3. Some common foot hazards include:
   a. Compression.
   b. Puncture.
   c. Extreme heat or cold.
   d. All the above.

4. What should your employer do first to reduce hand injuries?
   a. Provide gloves.
   b. Make sure all guards remain on equipment and tools.
   c. Implement engineering controls.
   d. Keep all irritating substances off the jobsite.

5. Hard hats should be:
   a. Inspected daily.
   b. Painted once a month.
   c. Stored in direct sunlight.
   d. Worn without the suspension.

6. Hearing protection can be:
   a. Plain cotton.
   b. Of the ear muff type.
   c. Ear plugs.
   d. Both b. and c.

7. Before you wear a respirator you must:
   a. Select the correct respirator for the task.
   b. Have a medical evaluation.
   c. Make sure the respirator fits properly.
   d. All the above.

8. Which of the following type of shoe or boots protects you against static electricity build up?
   a. Safety boots.
   b. Puncture resistant footwear.
   c. Conductive footwear.
   d. Open toed shoes.

9. Eye injuries can result from:
   a. Poor ventilation.
   b. Removing machine guards.
   c. Wearing eye protection.
   d. Both a. and b.

10. Eyewash facilities must be within _____ of the hazardous work area.
    a. 50 feet.
    b. 100 feet or a 10-second walk.
    c. 75 feet or a 10-second walk.
    d. 50 feet or a 10-second walk.

SCAFFOLDS:
REACHING SAFE HEIGHTS

A crew laying bricks on the upper floor of a three-story building built a six-foot platform spanning a gap between two scaffolds. The platform was correctly constructed of two 2" × 12" planks with standard guardrails; however, one of the planks was not scaffold grade lumber and also had extensive dry rot in the center. When a bricklayer stepped on the plank it disintegrated and he fell 30 feet to his death.

Two employees were sandblasting a 110 foot water tank while working on a two-point suspension scaffold 60-70 feet above the ground. The scaffold attachment point failed, releasing the scaffold cables, and the scaffold fell to the ground. The employees were not tied off independently and the scaffold was not equipped with an independent attachment system.

Cases like these remind us that the danger of working on scaffolding is very real. Despite OSHA's 1971 scaffold standard, 87 construction workers died from falls from scaffolding in 2002. That same year 3,853 workers were injured in scaffold or scaffold-related accidents. Of those injured, most said the accidents were caused by:

- Planks or supports giving way (the most common cause).

- Employees slipping on the scaffold.

- Being struck falling objects.

Where Are the Regulations?

The scaffold regulations are found at 29 CFR 1926.450-.454. The rules are divided into five sections:

- **Scope and application**—The rule applies to all scaffolds used in construction, alteration, repair (including painting and decorating), and demolition.

- **General requirements**—Requirements for capacity, construction, access, use, fall protection, and falling object protection when working on scaffolds.

- **Additional requirements**—Pinpoints specific types of scaffolds in use and applies additional requirements for working safely with them.

- **Aerial lifts**—Includes safety requirements for extensible boom platforms, aerial ladders, articulating boom platforms, vertical towers, and a combination of any such devices.

- **Training**—Provides specific training requirements for: employees who work on scaffolds and employees who assemble, disassemble, move, operate, repair, maintain, or inspect scaffolds. Retraining is covered.

This chapter will concentrate on safe work practices for those who use scaffolds daily. However, the scaffold rule is also packed with information and requirements for:

- Engineers who design and build scaffolds,

- Your company's competent person, and

- Scaffold assemblers and disassemblers.

General Requirements

You need to be able to recognize hazards associated with the type of scaffold you are using and know what to do when things don't seem to be as they should. The following rules will help you understand and recognize some of the things you should be aware of and looking for.

Working Platform Construction

- Platforms, on all working levels, must be fully decked between the front uprights and the guardrail supports.

- The space between planks, and the platform and uprights, can be no more than one inch wide. Exceptions are made when your employer can show that a wider space is necessary.

- Platforms and walkways must be at least 18 inches wide except that ladder jack, top plate bracket, and pump jack scaffolds must be at least 12 inches wide.

- If work areas are so narrow that platforms and walkways can't be 18 inches wide, they must be as wide as possible. You must also be protected from falls by guardrails and/or personal fall arrest systems.

- The front edge of platforms must not be more than 14 inches from the face of your work unless guardrails are erected along the front edge and/or a personal fall arrest system is being used. The maximum distance from the face for plastering and lathing work is 18 inches.

- The ends of your platform, unless cleated or somehow restrained, must extend over the center line of its support at least six inches, except each end of a platform: (1) 10 feet or less in length must not extend over its support more than 12 inches, or (2) greater than 10 feet in length must not extend over its support more than 18 inches, unless it:

  - Is designed to support workers and/or materials without tipping; or

  - Has guardrails to block employee access to the platform end.

Supported Scaffolds

- For every four feet a scaffold is high, it must be at least one foot wide. If it is not, it must be protected from tipping by tying, bracing, or guying per the OSHA rules.

- Supported scaffolds must sit on base plates and mud sills or other steady foundations.

- Objects, such as blocks of wood or buckets, must not be used to support scaffolds or be used as working platforms.

- Supported scaffold poles, legs, posts, frames, and uprights must be plumb and braced to prevent swaying and movement.

Suspension Scaffolds

- The inboard ends of suspension scaffold outriggers must be stabilized by bolts or other direct connections to the floor or roof deck or stabilized by counterweights.

- Your company competent person must check the connections before you use a suspension scaffold.

- Counterweights must be secured by mechanical means to the outrigger beams. They cannot be made of flowable material such as sand or gravel, or construction materials such as masonry units or rolls of roofing felt.

- Suspension ropes must be inspected by your competent person prior to each workshift and after every occurrence which could affect a rope's integrity.

- Report any of these "rope" problems to your supervisor:

  - Any physical damage which doesn't allow the rope to do what it is suppose to do or makes it weaker.

  - Kinks that might cause a problem during tracking or wrapping around the drum.

  - Broken wire strands.

  - Abrasions, corrosion, scrubbing, flattening or peening causing loss of more than one-third of the original diameter of the outside wires.

  - Evidence that the secondary brake was activated during an overspeed condition and has engaged the suspension rope.

- Gasoline-powered equipment and hoists must not be used on suspension scaffolds.

- Gears and brakes of power-operated hoists used on suspension scaffolds must be enclosed to prevent pinch hazards.

- Two- and multi-point suspension scaffolds must be tied or some other way secured to prevent them from swaying when necessary as determined by your competent person. Window cleaners' anchors can not be used for this purpose.

Access to Your Scaffold

Getting to the work level of a scaffold has always been a serious problem. When not provided with a proper stairway or ladder, you might be tempted to use crossbraces to climb the scaffold. *Never use crossbraces to gain access to a scaffold working platform.*

Access to and between scaffold platforms more than two feet above or below the point of access must be made by:

- Portable ladders, hook-on ladders, attachable ladders, scaffold stairways, stairway-type ladders (such as ladder stands), ramps, walkways, integral prefabricated scaffold access, or equivalent means; or

- By direct access from another scaffold, structure, personnel hoist, or similar surface.

Portable, Hook-On, and Attachable Ladders must be:

- Positioned so as not to tip the scaffold,

- Positioned so the bottom rung is not more than 24 inches above your starting point, and

- Equipped with a rest platform at 35-foot maximum vertical intervals.

Stairway-Type Ladders must:

- Be provided with rest platforms at 12 foot intervals, and

- Have slip-resistant treads on all steps and landings.

Stairtowers must:

- Be equipped with a stairrail consisting of a toprail (handrail) and a midrail on each side of each scaffold stairway,

- Have handrails surfaced to prevent injury from punctures, lacerations, and snagging of clothing,

- Have slip-resistant surfaces on treads and landings, and

- Have guardrails on the open sides and ends of each landing.

Ramps and Walkways six feet or more above lower levels must have guardrail systems in place.

Scaffold Frames that are designed to be used as access ladders must:

- Be designed and constructed for use as ladder rungs,

- Be uniformly spaced within each frame section, and

- Have a maximum spacing between rungs of 16 $^3/_4$ inches.

Using Scaffolds

Once you are ready to work on your scaffold you have a greater task ahead, concentrating on your work while ensuring you and your coworkers work safely. This is no time to become lax or take shortcuts. It can be a long way to the ground.

Before each work shift, and after any incident which could affect a scaffold's structure, your company competent person must inspect all scaffolds and scaffold components for visible defects.

When you get to your working surface take time to look around and ensure everything is as it should be. Some things you should look for when working from a scaffold include:

- The scaffold is not loaded more than its maximum intended load or rated capacity, whichever is less. Know this rating and know how to estimate the load (workers, tools, paint buckets, etc.) on the scaffold.

- The scaffold is not a shore or lean-to scaffold.

- Your scaffold is not moved horizontally while you are on it unless it has been designed to do that.

- The scaffold is far enough from power lines as follows:

Insulated lines

Voltage	Minimum distance	Alternatives
Less than 300 volt	3 feet	
300 volts to 50kV	10 feet	
More than 50 kV	10 feet + 0.4 inches for each 1 kV over 50 kV.	2 times the length of the line insulator, but never less than 10 feet

Uninsulated lines

Voltage	Minimum distance	Alternatives
Less than 50 kV	10 feet	
More than 50 kV	10 feet + 0.4 inches for each 1 kV over 50 kV.	2 times the length of the line insulator, but never less than 10 feet

Note: If you must be closer to power lines than specified above, consult the OSHA rules for proper procedures.

- Swinging loads being hoisted onto or near your scaffold have tag lines or other measures to control the load.

- You are not working on a scaffold during a storm or high winds unless your competent person says it is safe to do so and you are protected by a personal fall arrest system or wind screen.

- Debris is not allowed to accumulate on your scaffold.

- You are not using makeshift devices such as boxes and barrels to increase your working height.

- You are not working on a scaffold covered with snow, ice, or other slippery material except to remove the material.

- You do not increase your working height with a ladder except on large area scaffolds and you meet a list of requirements found in §1926.452(f)(15).

Fall Protection

Fall hazards account for a high percentage of the injuries and deaths for scaffold users. When you are on a scaffold platform more than 10 feet above a lower level you must be protected from falling by some type of fall protection. The type of fall protection required depends on the type of scaffold you use:

Scaffold Type	Fall Protection Required
Boatswains' chair, catenary scaffold, float scaffold, needle beam scaffold, or ladder jack scaffold	Personal fall arrest system
Single-point or two-point adjustable suspension scaffold	Personal fall arrest and guardrail system
Chicken ladder	Personal fall arrest, guardrail, or grabline
Self-contained adjustable scaffold when the platform is supported by the frame	Guardrail system
Self-contained adjustable scaffold when the platform is supported by ropes	Personal fall arrest and guardrail system
Walkway within a scaffold	Guardrail system within $9^1/_2$ inches of and along at least one side of the walkway
Overhand bricklaying from a supported scaffold	Personal fall arrest or guardrail system
All others	Personal fall arrest or guardrail system

All personal fall arrest systems used on scaffolds must meet the requirements of §1926.502(d) and must be attached by lanyard to a vertical or horizontal lifeline or scaffold structural member. When used, vertical and horizontal lifelines and lanyards must meet the requirements of §1926.452(g)(3).

When selected for fall protection, guardrail systems must:

- Be installed along all open sides and ends of platforms,

- Be installed before you can use it, and

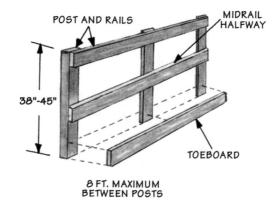

POST AND RAILS

MIDRAIL
HALFWAY

38"-45"

TOEBOARD

8 FT. MAXIMUM
BETWEEN POSTS

- Meet the physical requirements of §1926.451(g)(4).

  Note: Guardrail toprails manufactured or placed into service after January 1, 2000 must be between 38 and 45 inches high.

  Before this date, supported scaffold toprails, and all suspended scaffold toprails where a guardrail and personal fall arrest system is needed, can be between 36 and 45 inches.

- Screens and mesh, when used, must extend from the top edge of the guardrail system to the platform, along the entire opening between the supports.

- Steel or plastic banding cannot be used as a toprail or midrail.

- Manila or plastic rope used for toprails or midrails must be inspected by your competent person as often as necessary to ensure that it continues to meet the OSHA strength requirements.

- Crossbracing is acceptable in place of a midrail, but it must meet the requirements of §1926.452(g)(4)(xv).

Falling Object Protection

While working on scaffolding, you must be provided with not only hard hats, but additional protection from falling hand tools, debris, and other objects *above you*. This is usually done by installing equipment that contains or deflects the objects such as:

- Toeboards, screens, or guardrail systems; or

- Debris nets, catch platforms, or canopy structures.

When the objects are too large or heavy for the above-listed measures to work, workers above you must secure the objects away from the edge of the surface from which they could fall.

Where there is a danger of tools, materials, or equipment falling from a scaffold and striking fellow employees *below you*, your employer must use barricades or toeboards to protect them.

Where tools, materials, or equipment are piled higher than the top of the toeboard, paneling, screening, a guardrail system with openings small enough to prevent passage of objects, or canopy structures, debris nets, or catch platforms must be erected.

Specific Types of Scaffolds

In addition to the general requirements for scaffolds, the scaffold rule has additional information and safety requirements for particular types of scaffolds and features.

If you work on any of the following scaffolds, you — and your company competent person — need to review the safety requirements for that scaffold at 29 CFR 1926.452:

Pole Tube and coupler

Fabricated frame

Plasterers', decorators', and large area

Bricklayers' square

Horse

Form, and carpenters' bracket

Roof bracket

Outrigger

Pump jack

Ladder jack

Window jack

Crawling board

Step, platform, and trestle ladder

Single-point adjustable suspension

Two-point adjustable suspension (swing stages)

Multi-point, stonesetters' multi-point, and masons' multi-point adjustable suspension

Catenary

Float (ship)

Interior hung

Needle beam

Multi-level suspension

Mobile

Repair bracket

Stilts

Aerial Lifts

The scaffold rule contains safety requirements for aerial lifts because elevating and rotating work platforms are classified as scaffolds. The following types of aerial lifts, as well as a combination of any of these, are included in the rule:

Aerial ladders
Extensible boom platforms
Articulating boom platforms
Vertical towers

Some of the safety rules you must be aware of when working with or on such equipment are:

Aerial Ladders

Aerial ladders like ladder trucks and tower trucks must be secured in the lower position by the locking device on top of the truck cab, and the manually operated device at the base of the ladder before the truck is moved for highway travel.

Extensible and Articulating Boom Platforms

• Do not operate a boom platform unless you are authorized;

• Do not exceed boom and basket load limits, specified by the manufacturer;

• Have lift controls tested each day prior to use to determine if they are in safe working order;

• Before using a boom platform, set the brakes, place the outriggers on pads or a solid surface, and install wheel chocks if on an incline;

• Climbers must not be worn while working from boom platforms;

• To keep from falling, wear a body belt and a lanyard attached to the boom or basket (considered a restraining device), or a full body harness (considered a fall arrest device) attached to the boom or basket. Use only locking type snaphooks.

• Do not tie-off to a nearby pole, structure, or equipment when you are working from an aerial lift.

• Always stand firmly on the floor of the basket, and do not sit or climb on the edge or use planks, ladders, or other devices for a work position.

• Do not move a truck when the boom is elevated and workers are in the basket unless designed for this type of operation.

When extensible and articulating boom platforms are designed primarily as personnel carriers, they must have both platform and lower controls. Upper platform controls must:

- Be in or beside the platform within easy reach of the operator.

- Be plainly marked as to their function.

Lower controls must:

- Be plainly marked as to their function.

- Provide for overriding the upper controls.

- Must not be operated unless permission is obtained from the worker in the lift, *except in an emergency.*

The insulated portion of a boom platform must not be altered in any manner that might reduce its insulating value and pose an electrical hazard.

Review page 21 of this handbook for information on working near overhead power lines.

Training Requirements

Prior to working from any of the types of scaffolds, you must be trained to recognize the hazards associated with the type of scaffold you are using, and understand the procedures to control or minimize those hazards.

The training must include:

- The nature of any electrical, fall, and falling object hazards in your work area;

- The correct procedures for dealing with electrical hazards;

- The correct procedures for erecting, maintaining, and disassembling the fall protection and falling object protection systems being used;

- The proper use of the scaffold you are going to use;

- The proper handling of materials on that scaffold to include the maximum intended load and the load-carrying capacities of that scaffold; and

- Any other requirements of the scaffolding rule that applies to your work situation.

If you are erecting, disassembling, moving, operating, repairing, maintaining, or inspecting a scaffold, there are additional training requirements in the scaffold rule at 29 CFR 1926.454(b).

You must be retrained:

- Where changes at your worksite present a new hazard you have not been previously trained to recognize and deal with;

- Where changes in the types of scaffold, fall protection, or other equipment presents a hazard you have not been previously trained to recognize and deal with; and

- When your employer believes you lack the skill or understanding needed to work safely.

Work at Working Safely

The scaffold regulation, if followed, is designed to prevent accidents. Of course, no amount of rules or training can make you work safe. The desire to be as healthy when you leave your job as when you arrived is the best motivator to follow safe work habits.

NOTES

SCAFFOLDS REVIEW

1. The space between scaffold planks and the platform uprights can be no more than _____ wide.
   a. Three inches.
   b. Two inches.
   c. One inch.
   d. None of the above.

2. Unless guardrails or personal fall arrest systems are used, the front edge of platforms must not be more than _____inches from the face of your work. For plastering and lathing the maximum distance is _____ inches.
   a. 10, 14.
   b. 14, 18.
   c. 16, 20.
   d. 20, 24.

3. For every four feet a supported scaffold is high, it must be at least _____ wide.
   a. One foot.
   b. Two feet.
   c. Three feet.
   d. Four feet.

4. Your company's _____ must check the scaffold connections before you use a suspension scaffold.
   a. Authorized person.
   b. Vice president.
   c. Safety manager.
   d. Competent person.

5. Aerial lifts are not considered scaffolds.
   a. True.
   b. False.

6. Access to and between scaffold platforms more than two feet above or below the point of access must be made by:
   a. A ladder, stairway, ramp, or walkway.
   b. Direct access from another scaffold, structure, or personnel hoist.
   c. Both a. and b.
   d. Neither a. and b.

7. When you are on a scaffold platform more than _____ feet above a lower level you must be protected from falling by some type of fall protection.
   a. 6.
   b. 10.
   c. 13.
   d. 15.

8. The minimum distance your scaffold should be from an insulated electrical power line of less than 300 volts is:
   a. 10 feet.
   b. Six feet.
   c. Three feet.
   d. One foot.

9. Examples of equipment that contains or deflects falling objects are:
   a. Suspension ropes.
   b. Portable ladders.
   c. Supported scaffold poles.
   d. Debris nets, catch platforms, or canopy structures.

10. Gasoline-powered equipment and hoists must not be used on suspension scaffolds.
    a. True.
    b. False.

SITE SAFETY & SECURITY: THE KEY TO PREVENTING ACCIDENTS & LOSSES

Good construction site safety and security not only protects your company's assets, it also protects your own personal safety, your tools, and job. This chapter will give you an overview of your role in keeping your construction site safe and secure.

OSHA is concerned with workplace safety and health. Jobsite security is an issue each individual company must deal with. However, because everything that a company does goes hand-in-hand, if your company is a safe place to work, it most likely is a secure place to work.

Jobsite Safety

Your employer is responsible for training you in the recognition and avoidance of unsafe conditions and of the regulations applicable to your workplace. It is also your employer's responsibility to frequently and regularly visit the jobsite and inspect for safe working conditions. However, you are also an important player in jobsite safety and security.

Housekeeping

Housekeeping issues are usually not at the top of your priority list. But, most people do not enjoy working in a mess, and for good reason. It is distracting, unsafe, unsanitary, and it is against OSHA regulations.

The OSHA rules (29 CFR 1926.25) clearly state that you must work in conditions that promote safety and health. This means "picked-up" worksites. Some of the OSHA requirements are:

- Keep work areas, passageways, and stairs in and around your project free from scrap lumber and form lumber with protruding nails.

- Remove garbage, combustible scrap, and debris at regular intervals during the day.

- Collect and separate waste, garbage, and flammable rags in containers provided by your employer. Make sure containers for garbage and other oily, flammable, or hazardous wastes have covers.

- Report unsafe situations and accidents to your job foreman immediately.

The simplest of tasks—placing trash where it belongs, coiling up extension cords when not in use, and stacking lumber out of the way—may seem unimportant and unnecessary, until someone gets hurt.

Good housekeeping boosts morale, promotes safety, and encourages overall professional work habits, and adds dollars to the bottom line.

Hazard Recognition

Recognition, avoidance, and prevention of unsafe conditions should allow you to better control or eliminate any hazards or other exposure to illness or injury. Pay attention to all items discussed during training. You and your coworkers' lives depend on your understanding of how to recognize, avoid, and prevent the hazards at your worksite.

Harmful Substances

Know the safe handling and use of all poisons, caustics, and other harmful substances. And know the potential hazards, personal hygiene, and personal protective measures required. It may mean you will need to wear proper gloves or a respirator. It may also mean you will need to wash your hands after using the substance. Be sure to read the label and material safety data sheet (MSDS) for any harmful substance you work with. To find out more about harmful substances, see the Hazard Communication and the Jobsite Exposures chapters.

Harmful Plants and Animals

Do you see poison ivy around your site? How about a hornets' nest? Working in the great outdoors can give you a feeling of being "one with nature." But nature can bite if you're not careful.

What are the harmful plants and animals at your jobsite? Here's a list of possibilities:

Poison ivy

Poison oak

Poison sumac

Ragweed

Sagebrush

Poisonous snakes (rattlesnakes, copper heads, water moccasins, & coral snakes)

Ticks

Poisonous spiders (black widow & brown recluse)

Bees/Hornets/Wasps

Mosquitoes

Fire ants

Scorpions

Dogs and cats

Foxes

Wolves

Bears

Wildcats

Coyotes

Skunks

Raccoons

Squirrels

Bats

Crocodiles

Sharks/Barracuda/

Jellyfish

Know how you can avoid coming into contact with these harmful plants and animals:

Thing	Location	Protection
Poisonous plants	Eastern, Southern, and Midwestern U.S. The toxic oil can be on anything that touches the plant (even tools and equipment).	Remember the phrase, "Leaves of three, let them be." Wear long pants and long sleeves, and, whenever possible, gloves and boots.
Insects	Bees/hornets/wasps—Nests & flowering plants Black widow spiders—Dark, damp places like woodpiles Brown recluse spiders—Dark, dry places in Southern U.S. Lyme disease-carrying ticks—Anywhere in the U.S., but mostly in WI, MN, and on the east and west coasts Many insects forage on the ground or in low lying foliage	Wear foot protection. Avoid swatting at a flying insect. Gently brush it aside or wait for it to go away. Avoid wearing sweet-smelling colognes, deodorants, or other personal hygiene products. Avoid wearing bright-colored clothing. Avoid eating in areas where there are insects; many are naturally attracted to food odors.
Animals	Shelters, woodpiles, and waste dumps	Do not approach any animal.
Snakes	Rocks and woodpiles	Wear tall leather boots.

During first aid training, you should learn how to treat any injuries caused by local plants and animals (also see the First Aid and Bloodborne Pathogens chapter). Here are some general treatment methods:

Injury	Treatment
Possible contact with poisonous plant	Wash all exposed areas with cold running water. The water will neutralize the toxic oil and keep it from spreading. Do not use soap; it may spread the oil. Wash all clothing outside before bringing it into your house. Handle it as little as possible until it is soaked. Wash all gear that may also be carrying the oil. If you develop a rash, avoid scratching the blisters. Fingernails may carry germs and cause an infection.

Injury	Treatment
Insect bite or sting	Remove stinger by scraping it. You can use tweezers, but squeezing the stinger releases more venom.
	Apply hydrocortisone cream or calamine lotion.
	Apply cold water or ice for pain.
	Move indoors to avoid attracting other wasps (venom from wasp stings has an odor that attracts other wasps).
Animal bite or scratch	Clean wound with soap and water to prevent infection.
	Bandage wound.
Snake bite	Allow bite to bleed for 15 - 30 seconds.
	Clean and disinfect the area.
	Elastic bandage the area, but leave the bite marks uncovered.
	Press on the bite with a gauze pad and tape in place.
	Cool the wound without the use of ice.
	Seek medical attention.

Always seek medical attention if the wound becomes infected or does not improve, other symptoms like headache or fever develop, or poisonous plants and animals are involved.

Flammable Liquids, Gases, or Toxic Materials

Do you know how to safely handle and use flammable liquids, gases, or toxic materials? You should. See the Emergency Response, Hazard Communication, and Jobsite Exposures chapters. Also be aware of any environmental control measures you need to use.

What is the nature of the hazards involved in your confined space entry tasks? If the space is a permit-required confined space, these hazards should be listed on the permit. Find out what precautions you need to take to combat these hazards *before* you enter. And use all protective and emergency equipment required. See the Confined Space Entry chapter for more information.

Personal Safety Measures

General safety rules that can help eliminate injury are:

- Wear the appropriate clothing for the job you are doing. Cold and heat can both present problems. Shirts should be worn at all times.

- Wear appropriate safety shoes for the job you are doing.

- Personal protective gear must always be a part of your wardrobe when appropriate. Safety glasses, hard hats, and safety shoes should be a part of your permanent equipment and worn when required.

- Rings, watches, and other jewelry are always a hazard and should not be worn when working.

- Combustible materials are a fire hazard at any construction site. Smoking, portable heating equipment, small engines, and welding pose the greatest danger.

- Never participate in or tolerate horseplay at your construction site. Supposed fun can turn into tragedy.

Materials Handling and Storage

Storing materials properly takes constant effort. Some of the OSHA rules you need to keep in mind while storing materials at your storage yard or jobsite are found at the end of the Materials Handling and Storage chapter.

Signs, Tags, Barricades, and Signals

Working in and around construction sites is dangerous business. Using the proper signs, signals, and barricades can make your worksite safe by warning you and your coworkers of dangers. The OSHA rules for signs, tags, barricades, and signals are found in 29 CFR 1926.200-.203.

Signs and Tags

When you see warning signs at your construction site you should know what they mean and why they are there. Your safety depends on it. Some of the signs you may see and what they mean are:

- DANGER signs warn you of immediate hazards. These signs have a red upper panel, white lower panel, and black outline on the edges.

- CAUTION signs warn of potential hazards or against unsafe practices. Yellow and black are the prominent color of caution signs.

- EXIT signs point the way to safety. Red letters on a white background are required.

- TRAFFIC signs control traffic.

- ACCIDENT PREVENTION TAGS are used as a temporary means of warning you of an existing hazard, such as defective tools and equipment. These tags have a white or yellow background with red or black letters. They typically state "Do Not Operate," "Danger," "Caution," "Out of Order," or "Do Not Use."

Signs and symbols must be visible at all times when work is being done and must be removed or covered when the hazards no longer exist.

Barricades

A barricade is an obstruction to keep persons or vehicles from entering.

Signals

Signals are moving signs held by flaggers or devices like flashing lights to warn of hazards. When signs, signals, and barricades do not provide the necessary protection for operations on highways or streets, flagmen or other appropriate traffic controls must be provided. See the Work Zone Safety chapter for information about flagging.

Jobsite Security

Jobsite security refers to all the efforts to keep a facility safe and secure. Recently, it often refers to preventing a terrorist attack. Your company's jobsite is vulnerable because of the many different trades and workers coming and going. Depending on the nature of the job, there can also be a risk of theft of chemicals and explosives.

Jobsite security is important for many reasons. The primary reasons are to protect your employer's investment and equipment and to protect your own personal equipment. Construction equipment is very expensive and not just a pocket change expense. You also want to protect your site from the kids who love to climb on cranes and dozers.

What Can Your Company Do?

Efforts to maintain jobsite security include everything from controlling access by vehicles and pedestrians to using lighting and motion detection alarms.

Control Access

Your company can use measures to direct and restrict access to a jobsite or facility. For a jobsite without any buildings, these control measures can include fences, gates, and earthen berms. If you are working in an existing facility, the control measures should already be in place. They may consist of walls, fences, bushes, gates, and traffic islands.

Secure the Perimeter

Perimeter security refers to measures used to protect the approaches and access to the grounds, including adequate illuminations and signs detailing access rights and entrance locations. At a building renovation job, access to pedestrian entrances can be controlled through the use of security personnel, gates or turnstiles, locked keycard gates or doors, and surveillance cameras.

Control Vehicle Traffic

Vehicle traffic can be controlled through landscaping, gates, and manned security points. Contractors and contract employees can be required to park their vehicles in clearly marked locations away from any sensitive areas.

Screen Deliveries

Your location may establish a central delivery point for packages, equipment, materials, and supplies coming onto the jobsite. This provides a screening function to protect against things like bombs or biological hazards.

Use Security Staff

Some jobsites have uniformed security staff posted at strategic locations or patrolling the site. They can monitor security-sensitive or unoccupied areas of the facility and watch for fires, break-ins, environmental problems, or unauthorized access. This is especially important at night or on weekends.

Explosives Theft

Mine sites, Army depots, and even police bunkers are targets of thieves attempting to steal explosives. Some construction companies, and probably all demolition firms, store and use explosives. The OSHA regulations explain how to do that and also require your employer to report thefts to the Bureau of Alcohol, Tobacco, and Firearms (BATF) if they notice explosives are missing.

Some specific requirements for safeguarding explosives include:

- Account for all explosives at all time;

- Keep explosives not being used in a locked magazine;

- Maintain an inventory and use record of all explosives;

- Inspect storage facilities at least weekly; and

- Report loss, theft, or unauthorized entry into a magazine.

What Can You Do?

Somehow you and your company must control the jobsite. How can you do this?

- Question strangers and report to your supervisor any strange vehicles or people hanging around during your shift.

- Use storage sheds, if provided, for your small tools, or take your tools home.

- Ensure heavy equipment is locked.

- Ask the local police if they would patrol your jobsite as frequently as they can.

- Ask nearby businesses or neighbors to report any out of the ordinary activity to you. Provide them with a toll free number to call.

- Create a company security patrol.

Jobsite security depends on your knowledge of safety and security protocols and procedures. It is an issue everyone must take seriously.

Work at Working Safely

Accidents that injure people or damage property are liability risks. The way to minimize these potential losses to your company is to apply all the things we've reviewed here.

NOTES

Employee _____

Instructor_____

Date _____

Company _____

SITE SAFETY & SECURITY REVIEW

1. Your employer is responsible for:
   a. Making sure you get to work on time.
   b. Training you in the recognition and avoidance of unsafe conditions.
   c. Providing you with clean work cloths.
   d. None of the above.

2. Recognition, avoidance, and _____ of unsafe conditions should allow you to better control or eliminate any hazards or other exposure to illness or injury.
   a. Acceptance.
   b. Elimination.
   c. Prevention.
   d. Promotion.

3. Remove a bee's stinger by:
   a. Pulling it out with a tweezers.
   b. Applying tape to the area and then pulling the tape off.
   c. Applying cold water to freeze it out.
   d. Scraping it out.

4. Danger signs warn you of _____.
   a. Immediate hazards.
   b. Potential hazards.
   c. Severe hazards.
   d. Unsafe practices.

5. Rings, watches, and other jewelry are always a hazard and should not be worn when working.
   a. True.
   b. False.

6. Caution signs warn you of potential hazards or _____.
   a. Severe hazards.
   b. Unsafe practices.
   c. Immediate hazards.
   d. None of the above.

SITE SAFETY   **215**

7. Controlling access to a jobsite can be done with:
   a. Fences.
   b. Gates.
   c. Bushes.
   d. All the above.

8. Which of the following can cause combustible materials to start on fire?
   a. Welding.
   b. Portable heating equipment.
   c. Smoking.
   d. All the above.

9. Make sure containers for garbage and other oily, flammable, or hazardous wastes _____.
   a. Have covers.
   b. Are emptied every week.
   c. Have plastic bags in them.
   d. Have ventilation holes.

10. Some things you can do to keep a facility safe and secure include:
    a. Screen deliveries.
    b. Control vehicle traffic.
    c. Restrict access.
    d. All the above.

SLIPS, TRIPS, & FALLS: ON THE JOB SAFETY BASICS

In this culture, falls are often not taken seriously. On TV and in cartoons spectacular falls are done for special effects or to make us laugh. A person falls but doesn't get a scratch.

In reality, falls are accidents which often cause injury and lost time. Falls can often be fatal. There were 490 construction worker deaths in 2002 from falls (not including falls from ladders, stairs, or scaffolds). In 1998 there were 33,400 injuries caused by falls at the jobsite. In the same year, slips and trips caused 5,200 (3 percent) of construction injuries.

Falls are costly accidents. In the past, workers involved in falls lost a median of 8.5 work days because of their accidents. Injuries from falls may include cuts, bruises, muscle sprains and strains, broken bones, and back injuries.

Physical Factors at Work in a Fall

It might seem that an accident due to a loss of balance is pretty uncomplicated. Actually, slips, trips, and falls involve three laws of science:

Friction is the resistance between things, such as between your shoes and the surface you walk on. Without it, you are likely to slip and fall. A good example is a slip on ice, where your shoes can't "grip" the surface, you lose traction and you fall.

Momentum is affected by the speed and size of the moving object. You've heard the expression, "The bigger they are, the harder they fall." Translate that to mean the more you weigh and the faster you are moving, the harder your fall will be if you should trip or slip.

Gravity is the force that pulls you to the ground once a fall is in process. If you lose your balance and begin to fall, you're going to hit the ground. Your body has automatic systems for keeping its balance. Your eyes, ears, and muscles all work to keep your body close to its natural center of balance. A fall is likely if your center of balance (sometimes called center of gravity) shifts too far and can't be restored to normal.

Slips

A slip is a loss of balance caused by too little friction between your feet and the surface you walk or work on. Loss of traction is the leading cause of workplace slips. Slips can be caused by constantly wet surfaces, spills or weather hazards like ice and snow. Slips are more likely to occur when you hurry or run, wear the wrong kind of shoes, or don't pay attention to where you're walking.

Follow these safety precautions in order to avoid a slip:

- Practice safe walking skills. If you must walk on wet surfaces, take short steps to keep your center of balance under you and point your feet slightly outward. Move slowly and pay attention to the surface you're walking on.

- Clean up spills right away. Whenever you see any kind of spill, clean it up yourself or report it to the appropriate person. Even minor spills can be very hazardous.

- Don't let grease accumulate at your jobsite. If grease is present on the job, be sure that it's cleaned up promptly.

- Be more cautious on smooth surfaces. Move slowly on new floors which have been waxed, and other very slippery surfaces.

Wearing the Right Shoes Helps

One of the best ways to help prevent slip, trip, and fall injuries is to increase friction between your shoes and the surfaces you walk on. The amount of traction a sole provides varies with the work surface.

For instance, shoes with neoprene soles can be used safely on most wet or dry work surfaces. However, they are not recommended for oil conditions. Crepe soles are best for rough concrete, either wet or dry, but are not suggested for tile, smooth concrete, or wood surfaces.

When selecting safety shoes, you have to determine what conditions and/or hazards you face most often on the job. For example, non-slip shoes and soles will also be useful when climbing ladders or scaffolds.

Other devices are available to increase traction on your shoes. Strap-on cleats can be fastened to soles for greater traction on ice. There are non-skid sandals and boots that slip over shoes and offer better traction on ice, oil, chemicals, and grease. If these devices are available, use them as the job requires.

Neat and Orderly Work Areas Help

OSHA requires that your workplace be kept clear of scrap lumber and other debris. Floors must be dry and free of protruding objects such as nails, splinters, holes, or loose boards.

Many slip accidents are caused by improper cleaning methods. The average person takes about 18,000 steps each day on a wide variety of floor surfaces. Floors can be treated with many finishes and you might have occasion to select products to be used in clean-up or maintenance.

Added Traction for Wet Floors

One way to avoid slips on frequently wet surfaces is to apply some type of abrasive that will increase traction. Epoxies and enamels that contain gritty compounds may be painted on concrete, wood, and metal. These products are especially useful for passageways, walkways, and ramps.

Some manufacturers offer a selection of strips and rolls of skid-resistant material that you can apply to stairs, gangplanks or other potentially hazardous walking surfaces. Rubber mats can be used as a permanent or temporary solution to slippery areas. Use them, for example, near entrances of buildings under construction where water or other materials can be carried in from outdoors.

Trips

Trips occur whenever your foot hits an object and you are moving with enough momentum to be thrown off balance. A trip can happen when your work area is cluttered, when lighting is poor, or when an area has loose footing. Trips are more likely to happen when you are in a hurry and not paying attention to where you're going. Remember these rules to avoid tripping:

- Make sure you can see where you're going. Carry only loads that you can see over.

- Keep work areas well lit. Turned-off lights and burned-out bulbs can interfere with your ability to see clearly. Don't fumble in the dark. Use a flashlight or extension light to make your walking area visible in unlighted areas.

- Keep your jobsite clean and don't clutter passageways or stairs. Store materials and tools in cabinets or specially assigned storage areas.

- Arrange equipment so that it doesn't interfere with walkways or pedestrian traffic in your area.

- Extension or power tool cords can be dangerous tripping hazards. Working areas must be kept clear of them. Tape them to wall supports or arrange them so that they won't be in the way for other workers. However, they must not be fastened with staples, hung from nails, or suspended by wire.

- Eliminate hazards due to loose footing on stairs, steps, and floors. Report loose stair treads or handrails. Uneven walking surfaces, broken pavement or loose floor boards can also catch a foot and cause a fall.

- Store gangplanks and ramps properly.

Falls

Falls occur whenever you move too far off your center of balance. Slips and trips often push you off your center of balance far enough to cause a fall, but there are many other ways to fall. They are also caused by makeshift ladders, misuse of ladders, accidents while climbing, and improper scaffolding use.

Most falls are slips or trips at ground level, but falls from greater heights pose a much higher risk of serious injury. Avoid falls of any kind with these safety measures:

- Don't jump. Lower yourself carefully from trucks or work stages.

- Check lighting. Make sure stairs and sites are properly lit.

- Repair or replace stairs or handrails that are loose or broken. If maintenance isn't your job, report these hazards to the proper personnel at your jobsite.

- Don't store tools or equipment on stairs or in passageways.

- Wear good shoes with the appropriate non-skid soles.

Stairs Can Be Dangerous

Another high-risk area for the construction worker is stairs. Loss of traction causes the highest number of stairway slipping and falling accidents and is usually due to water or other liquid on the steps. Because we use stairs so often, it's easy to forget that they can be hazardous. You can protect yourself from injury:

- Use handrails whenever possible. If you are carrying something and can't grip the rail, use extra caution.

- Don't run up or down stairs or jump from landing to landing.

- Don't carry a load that you can't see over.

- Report any unsafe conditions promptly. Maybe you can't control lighting or clutter on the stairway, but you can report them to your site supervisor.

- Report broken stair treads or handrails. Make sure that even temporary stairways are structurally sound and have handrails.

Work at Working Safely

Preventing slips, trips, and falls is a task that depends on many factors—most importantly—you. You might not be able to change your workplace, but you can recognize dangers, work to eliminate hazards, and use safety devices and equipment.

Employee _____

Instructor_____

Date _____

Company _____

SLIPS, TRIPS, AND FALLS REVIEW

1. Slips are more likely to occur when you:
   a. Hurry or run.
   b. Wear the wrong kind of footwear.
   c. Don't pay attention to where you're walking.
   d. All the above.

2. One of the best ways for you to help prevent slip, trip, and fall injuries is to:
   a. Clean up spills within 12 hours.
   b. Take long strides when walking on wet surfaces.
   c. Increase the friction between your shoes and the surfaces you walk on.
   d. Wear neoprene shoes when walking on oily surfaces.

3. When selecting safety shoes, you have to determine:
   a. How good they look with your work cloths.
   b. What conditions and/or hazards you face most often on the job.
   c. If your company will pay for them.
   d. If they are DOT approved.

4. OSHA requires that your workplace floors are:
   a. Dry.
   b. Free of protruding objects.
   c. Kept clear of scrap lumber and debris.
   d. All the above.

5. _____ increases traction on surfaces.
   a. Oil.
   b. Abrasive material.
   c. Grease.
   d. Wax.

6. To avoid tripping:
   a. Carry loads you can't see over.
   b. Store loads and material in walkways.
   c. Keep work areas well lit.
   d. Stack extension cords only two high when placed on the floor.

7. Falls occur when you move too far off your center of balance.
   a. True.
   b. False.

8. Falls are a result of:
   a. Slips.
   b. Trips.
   c. Improper scaffold use.
   d. All the above.

9. Falls are costly accidents. Workers lose an average of _____ work days because of fall accidents.
   a. 1.5
   b. 8.5
   c. 20
   d. 50

10. One thing you should never do is carry a load you can't see over.
    a. True.
    b. False.

TOOL SAFETY: STAYING OFF THE CUTTING EDGE

There's no doubt, tools, hand and powered, enable us to be more efficient and productive. Unfortunately, the power and efficiency also can pose serious risk of injury or death. Too often, tragedy occurs before steps are taken to search out and avoid or eliminate tool-related hazards.

Where Are the Regulations?

OSHA regulations mention tool safety in many locations:

Rule type	29 CFR	Name
Construction	1926.300-.307	Tools, Hand and Power
	1926.404	Electrical Safety—Wiring Design and Protection
	1926.702	Concrete and Masonry Construction—Equipment and Tools
	1926.951	Power Transmission and Distribution—Tools and Protective Equipment
General Industry	1910.211-.219	Machinery and Machine Guarding
	1910.241-.244	Hand and Portable-Powered Tools and Other Hand-Held Equipment

Hazard Recognition

In the process of removing or avoiding hazards, you must learn to recognize:

- The dangers associated with different types of tools.

- The safety precautions necessary to prevent accidents associated with those hazards.

Generally, hazards from tools can be put into two categories:

- Power sources for the tools. Most often, this is electricity but can be air or hydraulics.

- Dangers from the tool's action—cutting, shearing, drilling or debris resulting from those actions.

Some General Safety Rules

These general safety rules apply to all situations involving tools:

- Keep your work area well lit, dry, and clean.

- Maintain your tools. This includes proper sharpening, oiling and storage.

- Regularly inspect tools, cords, and accessories. This is an OSHA electrical safety requirement. See the Electrical Safety chapter for more information.

- Replace problem equipment immediately. Make repairs only if you are qualified.

- Use safety features such as three-prong electrical plugs, double-insulated tools, and safety switches. Keep machine guards in good repair and in place.

- Use personal protective equipment (PPE) such as safety glasses, respirators, and hearing protection.

- Dress right. Choose clothing that will not tangle in tools, and do not wear jewelry.

- Choose the right tool for the job. Use the right-sized tool.

- Around flammable substances, sparks produced by iron or steel hand tools can be a dangerous ignition source. Where this hazard exists, spark-resistant tools made from brass, plastic, aluminum, or wood will provide for safety.

Hand Tools

Hand tools include anything from axes to wrenches. The greatest hazards posed by hand tools is misuse and improper maintenance. Some examples of tool hazards are:

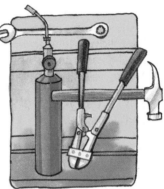

- Using a chisel as a screwdriver may cause the tip of the chisel to break, hitting you or other employees.

- If a wooden handle on a tool such as a hammer or an axe is loose, splintered, or cracked, the head may fly off and strike you or another worker.

- Using a wrench with sprung or smooth jaws.

- Impact tools such as chisels, wedges, or drift pins with mushroomed heads. The heads might shatter on impact, sending sharp fragments flying.

- Using a dull cutting tool.

Your employer is responsible for the safe condition of tools and equipment you use, but you have the responsibility to use and maintain tools properly. You should learn to use *all* tools at your jobsite—not just power tools—so that you understand potential hazards and the safety precautions needed to prevent those hazards from occurring.

For example, whether you're using a saw, knife, or other tools, direct them away from other employees working nearby. When working with draw knives, adzes, or similar cutting tools, use personal protective equipment such as wire mesh gloves, wrist guards, arm guards, and aprons or belly guards.

Power Tool Precautions

Power tools include electric, pneumatic, liquid fuel, hydraulic, or powder-actuated. Look at the following list to see what general precautions should be taken when using power tools:

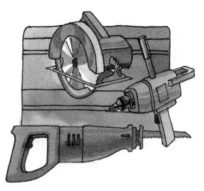

- Never carry a tool by the cord or hose.

- Never yank the cord or the hose to disconnect from the receptacle.

- Keep cords and hoses away from heat, oil, and sharp edges.

- Disconnect tools when not in use, before servicing, and when changing accessories such as blades, bits, and cutters.

- Keep all observers at a safe distance from the work area.

- Secure work with clamps or a vise, freeing both hands to operate the tool.

- Avoid accidental starting. Do not put your finger on the switch while carrying a plugged-in tool.

- Follow instructions in the user's manual for lubricating and changing accessories.

- Keep good footing and maintain good balance. This is another reason to keep your work area free from excess debris.

- Wear appropriate clothes. Loose clothing, ties, or jewelry can become caught in moving parts.

- Damaged portable electric tools must be removed from service immediately and tagged "Do Not Use." Ask your supervisor for approved company procedures for tagging broken equipment.

Guards

Hazardous moving parts—belts, gears, shafts, pulleys, sprockets, spindles, drums, fly wheels, chains, or other reciprocating, rotating, or moving parts of equipment—must be guarded if such parts are exposed.

Guards, as necessary, should be provided to protect the operator and others from:

- Point of operation hazards.

- In-running nip point hazards.

- Rotating parts.

- Flying chips and sparks.

Never remove safety guards when the tool is being used. For example, portable circular saws must be equipped with guards. An upper guard must cover the entire blade. A retractable lower guard must cover the teeth except when it makes contact with the work material. The lower guard must automatically return to the closed position when you're done making the cut.

Safety Switches

The following hand-held powered tools must be equipped with a momentary contact "on-off" control switch: drills, tappers, fastener drivers, grinders with wheels larger than two inches in diameter, disc sanders, belt sanders, reciprocating saws, saber saws, and other similar equipment. These tools may also be equipped with a lock-on control provided that turnoff can be accomplished by a single motion of the same finger or fingers that turn it on.

The following hand-held powered tools may be equipped with only a positive "on-off" control switch: platen sanders, grinders with wheels two inches or less in diameter, routers, planers, laminate trimmers, nibblers, shears, scroll saws, and jigsaws with blade shanks one-fourth inch wide or less.

Other hand-held powered tools such as circular saws, chain saws, and percussion tools without positive accessory holding means, must be equipped with a constant pressure switch that will shut off the power when the pressure is released.

Electric Tools

Among the chief hazards of electric-powered tools are burns and shocks which can lead to injuries or even heart failure.

To protect you, electric powered tools must either have a three-wire cord with ground and be grounded, be double insulated, or be powered by a low-voltage isolation transformer.

Three-wire cords contain two current-carrying conductors and a grounding conductor. One end of the grounding conductor connects to the tool's metal housing. The other end is grounded through a prong on the plug.

Anytime an adapter is used to accommodate a two-hole receptacle, the adapter wire must be attached to a known ground. Never remove the third prong from the plug.

Double insulation is more convenient. The user and the tools are protected in two ways: by normal insulation on the wires inside, and by a housing that cannot conduct electricity to the operator in the event of a malfunction.

You should follow these practices when using electric tools:

- Operate electric tools within their design limitations.

- Wear gloves and safety footwear.

- Store tools in a dry place when not in use.

- Do not use electric tools in damp or wet locations.

- Keep work areas well lighted.

Powered Abrasive Wheel Tools

Powered abrasive grinding, cutting, polishing, and wire buffing wheels create special safety problems because they may throw flying fragments.

Before an abrasive wheel is mounted, it should be inspected closely and sound- or ring-tested to be sure that it is free from cracks or defects. To test, tap wheels gently with a light non-metallic implement. If wheels sound cracked or "dead," they could fly apart in operation and so must not be used. A sound and undamaged wheel will give a clear metallic tone or "ring."

To prevent the wheel from cracking, be sure it fits freely on the spindle. The spindle nut must be tightened enough to hold the wheel in place, but not tight enough to distort the flange. Follow the manufacturer's recommendations. Assure that the spindle wheel will not exceed the abrasive wheel specifications.

Because a wheel might disintegrate (explode) during start-up, you should never stand directly in front of the wheel as it accelerates to full operating speed.

Portable grinding tools need to be equipped with safety guards to protect workers, not only from the moving wheel surface, but also from flying fragments in case of breakage.

In addition, when using a power grinder:

- Always use eye protection,

- Turn off the power when not in use, and

- Never clamp a hand-held grinder in a vise.

Pneumatic Tools

Pneumatic tools are powered by compressed air. They include chippers, drills, hammers, and sanders. The biggest danger in using pneumatic tools is getting hit by one of the tool's attachments or fasteners.

Pneumatic tools that shoot nails, rivets, or staples, and operate at more than 100 pounds per square inch, must be equipped with a special device to keep fasteners from being ejected unless the muzzle is pressed against the work surface.

Eye protection is required and face protection is recommended when using pneumatic tools. Safety shoes may also be required such as for jack hammers. Noise is another hazard. Working with noisy tools such as jackhammers requires proper, effective use of ear protection.

When using pneumatic tools, check that they are fastened securely to the hose. A short wire or positive locking device attaching the air hose to the tool will serve as an added safeguard.

Airless spray guns which atomize paints and fluids at high pressures (1,000 pounds or more per square inch) must be equipped with automatic or visual manual safety devices. These devices will prevent pulling the trigger until the safety device is manually released.

If an air hose is more than one-half inch in diameter, a safety excess flow valve must be installed at the source of the air supply to shut off the air automatically in case the hose breaks.

In general, the same precautions should be taken with an air hose that is recommended for electric cords. The hose is subject to the same kind of damage or accidental striking and presents tripping hazards.

A safety clip or retainer must be installed to prevent attachments, such as chisels on a chipping hammer, from being unintentionally shot from the barrel. Screens must be set up to protect nearby workers from flying fragments around chippers, riveting guns, staplers, or air drills. Compressed air guns should never be pointed toward anyone. Never "dead-end" it against yourself or anyone else.

Heavy jackhammers can cause fatigue and strains. Heavy rubber grips reduce these effects by providing a secure handhold.

Liquid-Fuel Tools

The most serious hazard with fuel-powered tools comes from fuel vapors that can burn or explode and give off dangerous exhaust fumes.

Handle, transport, and store gas or fuel only in approved flammable liquid containers, according to proper procedures for flammable liquids.

Before you refill the fuel tank, shut the engine down and allow it to cool to prevent igniting of hazardous vapors.

If you use a fuel-powered tool inside a closed area, effective ventilation and/or personal protective equipment is necessary to avoid breathing carbon monoxide. Fire extinguishers must be available in the area.

Powder-Actuated Tools

Powder-actuated tools operate like a loaded gun and should be treated with the same respect and precautions. In fact, they are so dangerous that they must be operated only by specially trained employees.

Safety precautions to remember are:

- Do not use in an explosive or flammable atmosphere.

- Before using the tool, inspect it. In particular, check that the barrel is free from obstructions.

- Never point it at anybody.

- Do not load until you are ready to use it. Do not leave a loaded tool unattended.

- Keep hands away from the barrel end.

To prevent the tool from firing accidentally, two separate motions are required for firing: one to bring the tool into position, and another to pull the trigger. The tools must not be able to operate until they are pressed against the work surface with a force of at least five pounds greater than the total weight of the tool.

If a powder-actuated tool misfires, wait at least 30 seconds, then try firing it again. If it still will not fire, wait another 30 seconds so that the faulty cartridge is less likely to explode, then carefully remove the load. The faulty cartridge should be put in water.

Use suitable eye and face protection when using a powder-actuated tool.

The muzzle end must have a protective shield or guard on the barrel to confine any flying fragments or particles when the tool is fired. The tool must be designed so that it will not fire unless it has this kind of safety device.

All powder-actuated tools must be designed for varying powder charges so that the user can select a powder level necessary to do the work without excessive force.

If the tool develops a defect during use, tag it and take it out of service until it is properly repaired.

Fasteners

Do not fire fasteners into material which would let them pass through to the other side. Also do not drive fasteners into very hard or brittle materials which might chip or splatter, or make the fastener ricochet. Fasteners must not be driven into materials like brick or concrete any closer than three inches to an edge or corner. In steel, fasteners must not come any closer than a half-inch from a corner or edge.

Use an alignment guide when shooting a fastener into an existing hole. A fastener must not be driven into a spalled area caused by an unsatisfactory fastening.

Hydraulic Power Tools

The fluid used in hydraulic power tools must be an approved fire-resistant fluid and must retain its operating characteristics at the most extreme temperatures to which it will be exposed.

Do not exceed manufacturer's recommended safe operating pressure for hoses, valves, pipes, filters, and other fittings.

Jacks

All jacks—lever and rachet jacks, screw jacks, and hydraulic jacks—must have a limiting device. Also, the manufacturer's load limit must be permanently marked in a prominent place on the jack.

Do not use a jack to support a lifted load. Once the load has been lifted, immediately block it. Use wooden blocking under the base if necessary to make the jack level and secure. If the lift surface is metal, place a one-inch thick hardwood block or equivalent between it and the metal jack head to reduce the danger of slippage.

To set up a jack, be sure that:

- The base rests on a firm level surface,

- The jack is correctly centered,

- The jack head bears against a level surface, and

- The lift force is applied squarely.

Proper maintenance of jacks is essential for safety. All jacks must be inspected before each use and lubricated regularly. If a jack is subjected to an abnormal load or shock, it should be thoroughly examined to make sure it has not been damaged.

Jacks exposed to freezing temperatures must be filled with an adequate antifreeze liquid.

Work at Working Safely

When using hand and power tools, you are exposed to a variety of hazards. But all hazards involved in the use of power tools can be prevented by following five basic safety rules:

- Keep all tools in good condition with regular maintenance,

- Use the right tool for the job,

- Examine each tool before use,

- Operate according to the manufacturer's instructions, and

- Provide and use the right protective equipment.

NOTES

TOOL SAFETY REVIEW

1. Tool power sources include:
   a. Electricity.
   b. Wind.
   c. Compressed air.
   d. Both a. and c.

2. Which of the following is *not* a general tool safety rule?
   a. Keep your work area well lit and clean.
   b. Maintain your tools properly.
   c. Carry tools by the cord or hose.
   d. Use personal protective equipment as needed.

3. Disconnect power tools when:
   a. Not in use.
   b. Servicing.
   c. Changing blades, bits, and cutters.
   d. All the above.

4. Do not put your finger on the _____ when carrying a plugged-in tool.
   a. Cord.
   b. On-off switch.
   c. Blade or bit.
   d. Hose.

5. Damaged portable electric tools must be:
   a. Used carefully.
   b. Put away immediately.
   c. Removed from service and tagged "Do Not Use."
   d. None of the above.

6. If a tool has a safety guard you must:
   a. Remove the guard just before using the tool.
   b. Never remove the safety guard when using the tool.
   c. Contact the tool manufacturer to see if the guard is really necessary.
   d. Carry the tool by the cord or hose.

7. Pneumatic tools are powered by:
   a. Electricity.
   b. Fuel (such as gasoline).
   c. Explosive powder.
   d. Compressed air.

8. Before an abrasive wheel is mounted, it should be inspected closely and:
   a. Prevented from cracking.
   b. Sound- or ring-tested.
   c. Not be dropped.
   d. Never held in a vise.

9. When using a hand-held power grinder:
   a. Always use eye protection.
   b. Turn off the power when not in use.
   c. Never clamp it in a vise.
   d. All the above.

10. If an air hose is more than one-half inch in diameter:
    a. A safety excess flow valve must be installed at the supply source.
    b. A safety clip or retainer must be installed.
    c. Both of the above.
    d. None of the above.

WELDING, CUTTING, AND BRAZING: AVOIDING THE 'TRIPLE THREAT'

Welding, cutting, and brazing poses unique threats to the health and safety of construction workers. Think about it — a welding arc is hot enough to melt steel, and the light it emits is literally blinding. It generates toxic fumes that are composed of microscopic particles of molten metal. Sparks and molten slag thrown by the arc can fly up to 35 feet and can cause fires.

Are you likely to be injured if you are a welder? Your job can be safe if you take the proper precautions and follow safe work practices. Even in small metal cutting jobs, try to resist the temptation to take shortcuts.

Where Are the Regulations?

The Occupational Safety and Health Administration (OSHA) has developed rules governing welding, cutting, and brazing. These regulations are found at 29 CFR 1926.350 through .354.

Welding Types

There are four different types of welding operations:

Welding type	Description
Oxygen-fuel gas	Joins metal parts by generating extremely high heat during combustion.
Resistance	Joins metals by generating heat through resistance created by the flow of electric current.
Arc	Joins or cuts metal parts by heat generated from an electric arc that extends between the welding electrode and the electrode placed on the equipment being welded. Includes gas-metal arc welding (also called metal inert gas welding) and flux-core arc welding (mistakenly called cored wire welding).
Other	This includes welding and cutting heat sources like friction, lasers, and ultrasonics.

Welding Hazards

As with any job activity, you will have hazards involved. However, welding operations expose you to a great number of hazards, minor to severe. Though welding hazards vary depending on the worksite and the job at hand, there are some common welding dangers you should be aware of, including:

- **Gases**—These may be released during welding and cutting operations. Gases can cause eye and respiratory irritation, headaches, coughing, dizziness, and chills, even death. Gases formed include carbon monoxide, nitrogen dioxide, ozone, and phosgene. These gases can form in many ways. Carbon monoxide, for example, can form if you use carbon dioxide shielding gas in gas metal arc welding.

- **Fumes**—These, too, are released during welding operations. The type of fume produced depends on the metal, metal preservatives, the electrode, or the filler rod used. A few hours of exposure to fumes can cause symptoms like headaches, muscle aches, general weakness, and chills. Fumes include:

Fume	Effects
Barium	Nose irritation, vomiting, heart trouble, muscle fatigue
Beryllium	Lung, liver, and kidney damage; breathing trouble
Cadmium	Lung and kidney damage, emphysema, chest pain, pulmonary edema (swelling)
Chromium	Skin, eye, and mucous membrane irritant; cancer
Copper	Vomiting, cold sweats, stiffness, chest pain, sinus burns
Fluoride	Eye, nose, and respiratory irritations; fluid in the lungs
Iron	Nose, throat, and lung irritation
Lead	Brain, kidney, muscle, nerve, circulation, and reproductive system damage; headaches; cramps
Magnesium	Nervous and respiratory system damage and irritation, eye irritation
Manganese	Trouble walking, weakness
Zinc	Vomiting, cold sweats, stiffness, chest pain, sinus burns

- **Radiation**—Looking at an arc without appropriate filter-plated eye protection can damage your eyes. Radiated heat can burn skin like a severe sunburn.

- **Electric shock**—Poorly grounded, defective, or improperly-operated equipment can kill you. Arc or resistance welding while standing on damp surfaces can create a prime situation for electric shock.

- **Fire and explosion**—Can be caused when welding and cutting near or on combustible or flammable materials, dust, vapor, liquid, or floors. Sparks fly into these places and can start a fire. Explosions can occur in areas with flammable liquids, gases, vapors, or dust.

- **Confined spaces**—Welding can displace oxygen and vapors can settle downward and even fill confined spaces. Flammable or combustible gases and toxic fumes could be present before and/or during entry. This can be deadly. Fire, explosion, and asphyxiation can occur.

- **Lead poisoning**—This hazard is created if you weld on lead-painted surfaces. Even seemingly well-ventilated areas can be hazardous.

- **Metal splatter, slag, and sparks**—These fly off the welding area and can hit you and/or burn you or your clothing.

- **Slips, trips, and falls**—Poor housekeeping in and around your welding area can lead to slips, trips, and falls. Working above the ground or floor level can create inherent falling hazards.

- **Noise**—This may seem like a minor hazard, but over time you could lose your hearing temporarily or permanently if you do not protect yourself from the noise produced by welding or cutting equipment.

- **Compressed gases**—These are hazardous because the gases are stored in high pressure. They can be flammable, poisonous, corrosive, or a combination. Flammable gases can ignite, explode if handled roughly or heated, flash back if vapors travel to an ignition source, or produce irritating or poisonous gas when burning. Gas and oxygen cylinders can even act like a rocket and break through concrete walls or travel through open spaces. Non-flammable gases can explode when mixed with fuels. Compressed gases can be harmful if inhaled, have extremely irritating vapors, and cause burns, dizziness, unconsciousness, or suffocation.

Protecting Yourself

While there are a lot of hazards, there are also a lot of ways to control or eliminate them to protect yourself:

- **Ventilation**—Exhaust hoods at the arc, fans, and open spaces all help to reduce the concentration of hazardous gases, fumes, and dusts. Ventilation also prevents the accumulation of flammable gases, vapors, and dusts that could cause fire. Know the symptoms of fumes and gases and get out of the area if they should develop. Perform atmospheric tests.

- **Avoid the plume**—Don't get too close to the gas or fume plume.

- **Personal protective equipment (PPE)**—This includes: flame-resistant aprons to protect against heat and sparks; leggings and high boots to protect you when doing heavy work; ankle-length safety shoes worn under your pant legs to keep from catching slag; shoulder cape and skull cap to protect against overhead welding; earplugs or ear muffs on very noisy jobs like those involving high velocity plasma torches; insulated gloves to protect against contact with hot items and radiation exposure; safety helmets to protect against sharp or falling objects; and goggles, helmets, and shields to protect your eyes and face. Use ANSI-approved filter lenses and plates to protect yourself from radiant energy. Also, protect onlookers by putting up shields so they won't be tempted to look at an arc.

- **Respirators**—When ventilation and plume avoidance don't give enough protection, or when welding creates an oxygen-deficient area, wear a respirator. Also, understand how to use your respirator.

- **Electrical precautions**—Do not arc weld while standing on damp surfaces or in damp clothing. Properly ground, install, and operate equipment. Do not use defective equipment. Use well-insulated electrode holders and cables. Insulate yourself from both the work and the metal electrode and holder. Don't wrap a welding cable around your body. Wear dry gloves and rubber-soled shoes. Do not use damaged or bare cables and connectors. In case of electric shock, do not touch a victim, turn off the current at the control box, and then call for help. After the power is off, you may perform cardiopulmonary resuscitation (CPR) if necessary.

- **Fire protection**—Wear flame-resistant clothing. Have someone be your fire watcher when you weld. Remove all combustible material at least 35 feet from the work area or move your work away from combustible materials. If neither of these options is possible, cover combustibles with fire resistant material. Do not weld in atmospheres containing dangerously reactive or flammable gases, vapors, liquid, or dust. Clean and purge containers which may have held combustible material before applying heat. Get a hot work permit and follow its safety precautions.

- **Confined spaces**—Limited work spaces, hazardous atmospheres, slippery floor surfaces, and even interior surfaces of the space should be evaluated for flammability, combustibility, or toxic fumes that could result from the welding process.

- **Clothing**—Wear wool, leather, or treated cotton clothing to reduce flammability for gas shielded arc welding. To avoid catching sparks, long sleeves and pants without cuffs/front pockets are recommended.

- **Lead**—Don't weld on lead-painted surfaces.

- **Fall protection**—Use a platform with railings, or safety harness and lifeline.

- **Compressed gases**—Handle cylinders carefully to prevent damage. Don't roll, drag, or slide cylinders. Don't lift them by their caps or use magnets to lift them. Use special hand trucks to lift cylinders in an upright position, or if the cylinders were manufactured with lifting attachments, then cradles or platforms may be used. Store cylinders in a safe, dry, well-ventilated area that is clean and free of combustible material. Avoid areas where cylinders can be knocked down or damaged.

Equipment Inspection and Maintenance

It almost goes without saying that welding equipment should be used according to the manufacturer's instructions. You must be familiar with the correct use and limitations of your welding equipment. In addition, routinely inspect and maintain your welding equipment, including welding cylinders. Inspect cylinders regularly to make sure all parts are in good working order, especially manifolds, headers, regulators, torches, and hose and hose connections.

Work at Working Safely

Getting the job done safely should always be your first concern. In all welding operations, take time to evaluate the job and implement appropriate safety precautions. This step will not only prevent equipment and machine damage, but it will reduce the risk of an accident that could injure you or a co-worker.

NOTES

WELDING, CUTTING, AND BRAZING

Employee _____

Instructor_____

Date _____

Company _____

WELDING, CUTTING, BRAZING REVIEW

1. Welding operations expose you to a few minor hazards.
    a. True.
    b. False.

2. Common welding hazards you could be exposed to include:
    a. Electric shock and radiation.
    b. Drowning
    c. Struck-by hazards.
    d. Trench cave-in.

3. Exposure to welding fumes can cause symptoms like:
    a. Skin rash.
    b. Hair loss.
    c. Headaches.
    d. None of the above.

4. Explosions caused by welding and cutting can occur in areas with:
    a. Flammable liquids.
    b. Gases and vapors.
    c. Dust.
    d. All the above.

5. What are some of the hazards of welding in confined spaces?
    a. Fire
    b. Explosion.
    c. Asphyxiation.
    d. All the above.

6. What types of clothing should be worn while welding:
   a. Clothing made of wool, leather, or treated cotton.
   b. Pants with cuffs.
   c. Pants with front pockets.
   d. Short pants.

7. Do not _____ while standing on damp surfaces.
   a. Resistance weld.
   b. Arc weld.
   c. Both the above.
   d. None of the above.

8. Remove combustible material at least ____ feet from the welding work area.
   a. 15.
   b. 20.
   c. 25.
   d. 35.

9. Compressed gas cylinders should be stored:
   a. In well-ventilated rooms.
   b. Near combustible materials.
   c. In a confined space.
   d. Near open flames.

10. Arc welding joins metal parts by generating extremely high heat during combustion.
    a. True.
    b. False.

WORK ZONE SAFETY: SAFETY IS NO ACCIDENT

In 2002, there where 117,567 work zone crashes and over 52,000 injuries — 1.8 percent of all roadway injuries. There was one work zone fatality every seven hours and one work zone injury every 15 minutes.

Highway construction workers are also in grave danger: they are twice as likely to be killed than all other construction workers combined. Work zone safety is no doubt one of the most important issues faced by the construction industry today.

Because most of our road-ways are in place, construction workers are not building many roads; they are repairing or expanding them. As our roadways age, more and more repair projects are needed. And as more motorists use existing roadways, road expansion projects are needed.

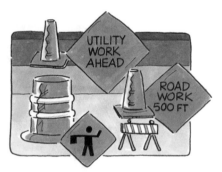

At the same time, due to pressure from motorists who do not want to be delayed by work zones:

• Today's motorists are allowed to keep high speeds through work zones,

• Work zones are shortened with less warning to motorists, and

• Roadwork is done at night where visibility is reduced.

Take precautions to protect you and your coworkers.

What Is a Work Zone?

A work zone is an area where roadwork is going on and traffic is affected. Construction or utility work done outside the roadway is not considered work zone activity. Examples of work zone activity include: building a new bridge; adding travel lanes to the roadway; extending an existing roadway; repairing potholes; and repairing electric, gas, or water lines within the roadway. Most work zones are divided into four areas:

1. **Advance warning area**—Section of the roadway where drivers are informed as to what's coming up.

2. **Transition area**—Section of the roadway where drivers are redirected to a temporary lane.

3. **Activity area**—Section of the roadway where work activity takes place. It includes both a **work space** where workers, equipment, and material is closed off to traffic, and a **traffic space** where traffic is routed through the activity area. The activity area may also contain unused **buffer spaces** to protect both workers and drivers.

4. **Termination area**—Section of the roadway used to return drivers to their normal lanes.

Where Are the Regulations?

OSHA rules for work zone safety can be found in 1926.200 to .203, Signs, Signals, and Barricades. Unfortunately, OSHA does not provide much in the way of protecting workers from the hazards of traffic in work zones.

However, while the Department of Transportation (DOT) has the intention of protecting motorists and pedestrians, it also provides some protection for road workers under: 23 CFR 630, Subpart J, Traffic Safety in Highway and Street Work Zones, and 23 CFR 655, Subpart F, Traffic Control Devices on Federal Aid and Other Streets and Highways.

Both OSHA and DOT refer to a national standard for traffic control on all public roads, including work zones. This standard is called the *Manual on Uniform Traffic Control Devices (MUTCD)*.

What Are the Hazards?

According to the Bureau of Labor Statistics, in 2002 there were 70 fatalities related to employees working in work zones. Some specifics are as follows:

- 59 fatalities caused by transportation incidents such as: collisions between vehicles or mobile equipment, workers struck by vehicle or mobile equipment in the roadway or on the side of the road

- 11 fatalities were caused by contact with objects and equipment other than vehicles or mobile equipment.

- Eight workers performing flagging or directing of traffic were killed.

- 18 employees were killed walking in or near the roadway.

- Vehicles were the primary source of death in 43 of the cases.

Here is a more complete list of work zone hazards you may face:

Being struck by motorists	Ungrounded electrical equipment	Vibrating tools
Being struck by heavy or rotating equipment	Poorly maintained tools and jacks	Heavy lifting
Heavy equipment rollover	Unguarded moving parts	Carbon monoxide from vehicle exhaust
Overhead power lines	Falling hazards (i.e., bridge sites)	Asphalt fumes
Underground electrical lines	Excavation cave-ins	Extreme heat/cold
		Rain and fog
		Darkness at night

While the public traveling through the work zone is important, our focus here is the safety of the worker performing tasks within the work zone. Let's take a look at the ways to keep you safe.

Traffic Control Plan

The Traffic Control Plan (TCP) describes the measures used to keep traffic moving safely and efficiently through the work zone. The measures selected depend on the type of roadway, the traffic condition, how long the project will be, location restrictions, and how close the work space is to traffic. A TCP can be very detailed and contain drawings of the particular work zone.

Your company may also create an internal traffic control plan to coordinate the flow of construction vehicles, equipment, and workers operating close to the activity area. When looking at the plan, workers on foot should pay attention to the areas they are not allowed to go into.

Signs

There are three types of traffic control signs:

- **Regulatory signs**—Inform roadway users of traffic laws. With some exceptions, these signs are rectangular and black and white. Examples include the STOP, YIELD, DO NOT ENTER, SPEED LIMIT, and ONE WAY signs.

- **Warning signs**—Notify drivers of conditions. With some exceptions, these signs are diamond-shaped and orange and black. Examples include the ROAD WORK, DETOUR, ROAD CLOSED, and RIGHT LANE CLOSED signs.

- **Guide signs**—Give information to help drivers with, for example, temporary routes, directions, and work being done. These signs are generally orange and black. Examples include the ROAD WORK NEXT # MILES, END ROAD WORK, and DETOUR ARROW signs.

These signs must be visible at all times when work is being done and must be removed or covered when the hazards no longer exist. At night, signs must be retroreflective or illuminated. If signs become worn or damaged, they must be replaced. Generally, signs should be placed on the right-hand side of the roadway.

Other signs you might find at your work zone which do not control traffic include: danger, caution, and exit signs and tags. See the Site Safety chapter for more information.

Signals

Signals warn of possible or existing hazards. They include:

- Sign paddles or flags held by flaggers,

- Portable changeable message boards, and

- Flashing arrow displays.

Channelizing Devices

Channelizing devices include, but are not limited to: cones, tubular markers, vertical panels, drums, barricades, temporary raised islands, and barriers. These devices protect workers in the work zone, warn and alert drivers to conditions created by roadwork, and guide drivers. Make sure these devices are clean and visible.

Flaggers

When signs, signals, and barricades do not provide enough protection for operations on highways or streets, then flaggers or other traffic controls must be provided. However, flagging is dangerous because it exposes the flagger to traffic. If you are a flagger, you should follow these rules:

* Use red flags (at least 18 inches square) or proper sign paddles when hand signaling in daylight. Most sign paddles have a red STOP sign on one side and an orange SLOW sign on the other. The MUTCD suggests that flags only be used in emergency situations.

* Use red lights when hand signaling at night.

* Use signals that conform to the MUTCD. Here are the signaling methods for sign paddles:

  * **To stop traffic**—Face traffic and hold the STOP sign paddle toward traffic with your arm extended horizontally away from the body. Raise your free arm with your palm toward approaching traffic.

  * **To direct stopped traffic to proceed**—Face traffic and hold the SLOW paddle toward traffic with your arm extended horizontally away from the body. Motion with your free hand for traffic to proceed.

- **To alert or slow traffic**—Face traffic holding the SLOW paddle toward traffic with your arm extended horizontally away from the body. You may motion up and down with your free hand, palm down, indicating that the vehicle needs to slow down.

- Wear a red or orange vest, shirt, or jacket.

- Wear white pants and a reflectorized vest, shirt, or jacket and a reflectorized hard hat at night.

- Coordinate with other flaggers and communicate by radio if you have no visual contact.

- Know how to combat both heat and cold exposure, dress appropriately, and know where shelter is available.

- Be alert to symptoms associated with carbon monoxide poisoning from vehicular traffic (nausea and headache). If symptoms develop, get to fresh air.

- Use barricades, cones, tubular markers, vertical panels, drums, and barriers to mark areas.

- Be aware of construction equipment around you. In order to know what is approaching from behind, you may need to wear a hard hat mounted mirror, have a buddy "spot" you, or use some kind of motion detector. Equipment operators, too, should know where you are.

Flagging can be a safe job if you remain alert to everything around you at all times.

Safe Work Practices

When working near traffic or heavy equipment:

- Wear highly visible clothing and a light-colored hard hat. During the day, you must wear a vest, shirt, or jacket that is orange, yellow, yellow-green, or a fluorescent version of these colors. At night, the vest, shirt, or jacket must be retroreflective. The retroreflective material must be orange, yellow, white, silver, strong yellow-green, or a fluorescent version of one of these colors and shall be visible at a minimum distance of 1,000 feet. Also, it is best to also wear white pants and a white reflectorized hard hat at night.

- Work where drivers can see you, but as far as possible from traffic. Be aware that drivers may not be able to see you when the sun is low in the sky or when it is rainy, foggy, or dark.

- Get in and out of traffic spaces and heavy equipment areas quickly and safely.

- Stay alert and don't wear a radio headset.

Do not operate equipment or a vehicle unless you are trained and authorized to operate that equipment. When operating equipment or vehicles:

- Always wear your seat belt when operating equipment or vehicles.

- Never move equipment without making visual contact with workers on foot near the equipment.

- Make sure equipment is inspected daily and that any problems are corrected. Report equipment problems.

- Use equipment with rollover protective structures.

- Chock two wheels when leaving equipment.

- If you must park your vehicle near traffic, park where drivers can see you (don't park around blind corners).

Other Work Zone Protective Measures

Other work zone safety precautions and protective measures could include:

- **Temporary barriers**—These devices prevent vehicles from entering areas where hazards, workers, or pedestrians may be.

- **Lower speeds**—If workers are especially vulnerable, work zone engineers should consider lowering the speed of traffic.

- **Shadow vehicle**—If roadwork is mobile, like for pothole patching, a vehicle with proper lights, signs, or a rear impact protector should be used to keep you from being hit.

- **Vehicle arresting systems**—This is fencing, cable, or energy absorbing anchors that prevent vehicles from entering activity areas while allowing the vehicle to safely slow down.

- **Rumble strips**—These consist of textured road surfaces that alert drivers to changing conditions.

- **Road closure**—If alternate routes can handle additional traffic, the road may be closed temporarily to give you the greatest protection.

- **Law enforcement**—If you are at high risk, police units may be placed to reduce traffic speeds.

- **Lighting**—To increase visibility at night, the work area and approaches should be well lit. However, floodlights must not create glare for drivers. Low-level truck lights also help operators see workers on foot. Reflective tape or light strips that outline the height and width of construction vehicles and equipment is beneficial.

- **Intrusion warning devices**—These devices may alert you of vehicles that accidentally enter the work space.

Training

If you are a work zone employee, you must be trained on all aspects of your job, including:

- Your role and location at the site,

- Traffic patterns and heavy equipment operations,

- Recognizing and eliminating or avoiding hazards,

- Understanding flagger signals and safety colors,

- Knowing communication methods and alarms,

- Knowing how to work next to traffic and heavy equipment in a way that minimizes accidents,

- Knowing your escape routes,

- Proper life saving personal protective equipment,

- Being as visible as possible, and

- Knowing how to operate equipment and vehicles and prevent rollovers.

Your company must familiarize you with these things before you work at each work zone and when a work zone changes.

Work at Working Safely

Working at work zones is dangerous but does not have to be unsafe. When company training is provided, and when you are on the jobsite, be alert. It could save your life.

Employee _____

Instructor_____

Date _____

Company _____

WORK ZONE SAFETY REVIEW

1. The portion of a work zone that informs drivers as to what's coming up is the:
   a. Termination area.
   b. Activity area.
   c. Transition area.
   d. Advance warning area.

2. Which of the following is a work zone hazard?
   a. Excavation cave-ins.
   b. Overhead power lines
   c. Darkness at night.
   d. All the above.

3. What is the name of the national standard for traffic control on public roads, including work zones?
   a. *Professional Road Workers Manual* (PRWM).
   b. *Manual on Uniform Traffic Control Devices* (MUTCD).
   c. *Street and Highway Construction Guide* (SHCG)
   d. *Public Road and Work Zone Manual* (PRWZM)

4. The area of the work zone where the work is being performed is called the:
   a. Advance warning area.
   b. Transition area.
   c. Activity area.
   d. Termination area.

5. Both OSHA and the _____ have rules on work zone safety.
   a. Department of Transportation.
   b. Department of Labor.
   c. Department of Homeland Security.
   d. Social Security Administration.

6. The _____ describes the measures used to keep traffic moving safely and efficiently through individual work zones.
   a. ANSI standard.
   b. Traffic Control Plan.
   c. MUTCD.
   d. ASME standard.

7. Which of the following is *not* a type of signal:
   a. Sign paddles or flags held by flaggers.
   b. Traffic control plan.
   c. Flashing arrow displays.
   d. Portable changeable message boards.

8. Flagging is dangerous because it exposes the flagger to _____
   a. Items falling off vehicles.
   b. Heavy lifting.
   c. Overhead power lines.
   d. Traffic.

9. When working near traffic or heavy equipment:
   a. Keep your back to oncoming traffic.
   b. Wear dark clothing during the day.
   c. Stay alert.
   d. Shout a warning to moving vehicles that come too close to you.

10. When operating equipment or vehicles:
    a. Move equipment without making visual contact with workers on foot nearby.
    b. Use equipment that doesn't have rollover protective structures.
    c. Park your equipment around blind corners.
    d. Make sure equipment is inspected daily.

NOTES

NOTES